ESSENTIALS OF
Modern Thermodynamics

Alejandro L. Garcia
San Jose State University

Contents

Preface

There can be no doubt, I think, that the average physicist is made a little uncomfortable by thermodynamics. - P.W. Bridgman, *The Nature of Thermodynamics*

As it says on the cover, this small book covers the essential elements of modern thermodynamics. In physics, as in art, the term *Modern* refers to something that is "only" a hundred years old. Classical Thermodynamics was mostly complete by the late 1880's and the theory of Modern Thermodynamics was developed mostly between 1900 and 1960. So Modern Thermodynamics is about the same "age" as Quantum Mechanics. Unlike quantum mechanics, modern thermodynamics is only recently appearing in undergraduate physics textbooks.* While classical thermodynamics focuses on systems at equilibrium, modern thermodynamics emphasizes irreversible processes in systems out of equilibrium. You could say it puts the "dynamics" back into thermodynamics.

This book is short by design so many topics are missing, with the most controversial omission being statistical mechanics. This was a deliberate choice since there are many excellent thermal physics books that cover this and other omitted topics, such as phase transitions. This book is also not designed as a course textbook although it should help you understand thermodynamics, in or out of school.

Reading Guide After reviewing the thermodynamics vocabulary defined in Chapter 1, if you're already comfortable with the First Law then jump right to Chapter 5. Or even Chapter 7 if you don't need a refresher on entropy in classical thermodynamics. I suggest skimming or skipping the optional sections and chapters, marked with an asterisk, on the first reading.

The first edition of *Numerical Methods for Physics* appeared nearly 30 years ago and, at the time, I swore that I would never write another book. This is that book; hope you like it.

Alejandro L. Garcia *San Jose, California*

---

*Modern thermodynamics does appear in chemistry textbooks, such as the excellent one by Kondepudi.

To Patrick Hamill
Friend, colleague, and the better writer

Chapter 1

Thermodynamic Systems

Much of this chapter is a summary of the vocabulary of thermodynamics. To provide context we'll consider several physical systems, describe how they can be modeled, and pose questions that can (or cannot) be answered by thermodynamics.

1.1   System #1: Gas-filled balloon

Figure 1.1 illustrates our first example: a volume of gas contained in a balloon. We'll focus on just the gas, treating the rubber shell as the boundary of the system. In thermodynamics this may be modeled as a *homogeneous* system with uniform temperature, T, and pressure, p. Other physical quantities include the volume, V, the mass M, the composition (e.g., 75% nitrogen and 25% oxygen), and possibly others. A *pure* system has only one type of molecule; if there are many types then the system is a *mixture*.

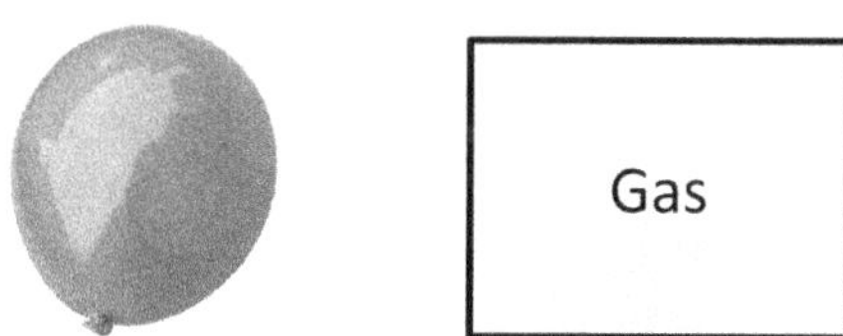

Figure 1.1: Gas-filled balloon: (left) physical system; (right) thermodynamic model.

The *state variables* are a set of quantities that define the thermodynamic state of our system. For a pure system (e.g., helium-filled balloon) these could be T, p, and M. Instead of mass we may want to use the number of molecules, $N = M/m$, where m is the mass of a molecule, or use the number

Extensive Variables		
M	System mass	kg
V	Volume	m^3
$\mathcal{N}$	Number of molecules	-
$N = \mathcal{N}/N_A$	Number of moles	mol
U	Internal energy	J
S	Entropy	J/K
H	Enthalpy	J
C	Heat capacity	J/K
Intensive Variables		
T	Temperature	K
p	Pressure	J/m^3
$\rho = M/V$	Mass density	kg/m^3
$n = \mathcal{N}/V$	Number density	m^{-3}
$[k] = N_k/V$	Concentration of species k	$M \equiv mol/l$
$m = M/\mathcal{N}$	Mass of a molecule	kg
$c = C/M$	Specific heat capacity	J/(K kg)
$\hat{c} = C/N$	Molar heat capacity	J/(K mol)
μ	Specific chemical potential	J/kg
$\hat{\mu}$	Molar chemical potential	J/mol
Constants		
$N_A = 6.02 \times 10^{23}$	Avogadro constant	mol^{-1}
$k_B = 1.38 \times 10^{-23}$	Boltzmann constant	J/K
$R = k_B N_A = 8.31$	Molar gas constant	J/(mol K)

Table 1.1: List of common thermodynamic variables and constants. A subscript k indicates species or type, such as m_k is the mass of molecules of species k.

of moles, $N = \mathcal{N}/N_A$, where N_A is the Avogadro constant (see Table 1.1). Other physical quantities may be expressed as functions of the chosen state variables, for example, volume $V(T, p, M)$.

Suppose that we have two identical balloons, that we treat as a single system. The larger system has the same temperature and pressure but double the volume and double the mass. Thermodynamic variables, such as p and T, that don't change when we duplicate like this are called *intensive* variables and those that double, such as V and M, are called *extensive* variables. We'll sometimes use intensive variables that are formulated from ratios of extensive variables, for example mass density is $\rho = M/V$.

For the gas in a balloon using thermodynamics we can answer questions such as:

- How does the volume vary with temperature and pressure?

- How much energy must be added to raise the temperature of the gas by one degree?

- Do the answers to these questions depend on the composition of the gas (e.g., pure helium versus a mixture such as air); what if the balloon is filled with a liquid (e.g., water) instead of a gas?

1.2   System #2: Cup of tea

Figure 1.2 illustrates our second example: brewed tea in a cup. Our thermodynamic model treats this as a *heterogeneous* system consisting of two subsystems: the liquid tea and the solid cup. Each subsystem is treated as homogeneous with its own state variables, such as the temperatures T_{tea} and T_{cup}.

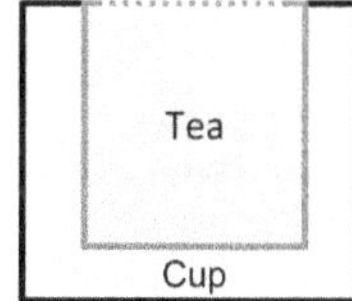

Figure 1.2: Cup of tea: (left) physical system; (right) thermodynamic model.

Thermodynamic systems are classified as (see Figure 1.3):

- **Homogeneous** All the intensive state variables (e.g., T, p) have the same value everywhere within the system.

- **Heterogenous** The system is divided into two or more parts and each part is homogeneous.

- **Continuous** The state variables may vary everywhere within the system.

An example of a continuous system would be an iron bar that's heated on one end and cooled at the other end.

For heterogeneous systems the two subsystems may interact with each other. For example, if the tea is hot and the cup is cold then we expect heat to flow between them. Similarly, water may evaporate from the hot tea, which is a flow of mass out of the system. We categorize thermodynamic systems according to their interactions (see Figure 1.4):

- **Isolated** There are no interactions; the total energy and matter in the system remains constant.

- **Closed** Energy may enter or leave the system but not matter.

- **Open** Both energy and matter may flow in or out.

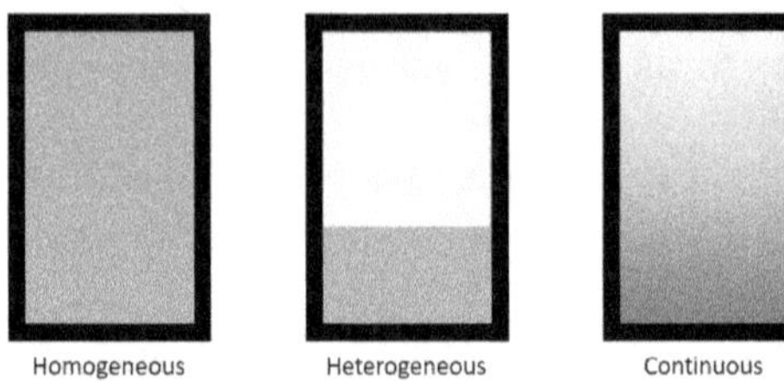

Figure 1.3: Illustrations of a homogeneous, a heterogeneous, and a continuous system.

In our cup of tea example if evaporation is negligible then the tea and the cup may each be treated as closed systems.

Interactions of a system with its environment or between subsystems occur at boundaries. Thermodynamic systems are modeled as being enclosed by boundary walls of different types, such as:

- **Adiabatic** No exchange of energy by heat occurs with an adiabatic wall.

- **Conductive (or Diathermal)** Exchange of energy by heat can occur with a conductive wall.

- **Stationary** No exchange of energy by compression work occurs with a stationary wall.

- **Mobile (or Piston)** Exchange of energy by compression work can occur with a mobile wall, which is usually thought of as a piston, when a force is exerted over a distance.

- **Solid** No exchange of matter occurs with a solid wall.

- **Permeable** Exchange of matter (and energy) does occur with a permeable wall. A semi-permeable wall allows only certain species to pass through it.

In Figure 1.4 the isolated system has solid, adiabatic, stationary walls. In that illustration the closed system has a mobile wall on top and a conductive wall on the bottom. When the boundary holds the temperature fixed we call say the system is *isothermal*. When a system is held at constant pressure (e.g., with a piston) we call it *isobaric*; if it is held at constant volume then it is *isochoric*. The open system in Figure 1.4 is enclosed by a permeable wall so both matter and energy are exchanged with the surroundings.

The closed system in Figure 1.4 exchanges energy with a *reservoir*. A reservoir subsystem is considered to be very large so the intensive variables (e.g., temperature) of the reservoir are taken to be fixed constants. Similarly, the open system in Figure 1.4 exchanges of both energy and matter

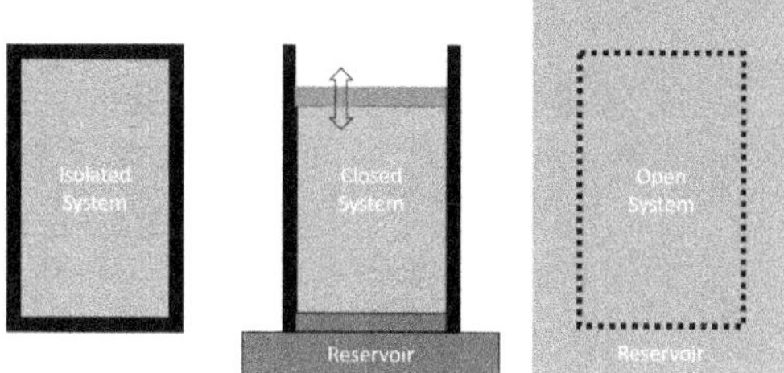

Figure 1.4: Illustrations of an isolated, a closed, and an open system.

but the reservoir's intensive variables (e.g., temperature and density) remain constant.

For heterogeneous systems, like this cup of tea example, thermodynamics answers questions such as:

- If the two subsystems are initially at different temperatures then experience indicates that they approach a common temperature. What is that temperature and how long does this take? How do these answers depend on the material properties (e.g. ceramic cup versus plastic cup)?

- If the tea is hot and the cup is cold then we expect energy to flow out of the tea. Can some of that energy be converted into a useful form, such as electricity? If yes then how much?

- Is it possible for energy to flow out of a cold cup and into the hot tea? Refrigerators transfer energy from cold to hot so why can't the cup be used to heat the tea?

Note that we will *not* address hydrodynamic questions, such as the velocity of water when stirred with a spoon.

1.3 System #3: Cat

Our final example is a domestic house cat (*Felis catus*) modeled as a continuous open system (Figure 1.5). Being a continuous system the intensive state variables, such as $T(\mathbf{x}, t)$, are functions of position $\mathbf{x}$ and time t. As with all living systems the cat is not a pure system but rather a mixture of many types of molecules. Furthermore, there are many chemical reactions occurring at the cellular level that are essential to biological life.

If a thermodynamic model includes transformations then the system is said to be *reactive*. For example there are chemical transformations, such as the reaction,

$$2\mathrm{H}_2\ (g) + \mathrm{O}_2\ (g) \rightleftharpoons 2\mathrm{H}_2\mathrm{O}\ (\ell),$$

Figure 1.5: Domestic cat: (left) physical system; (right) thermodynamic model.

which is hydrogen and oxygen gas forming liquid water. Since this is not a chemistry textbook we will write such transformations in a generic form,

$$2\mathbb{X} + \mathbb{Y} \rightleftharpoons 2\mathbb{Z}.$$

When a system has no transformations then it is *inert*. For an inert system the composition can only change if the system is open. The cat is an open, reactive system with a mixture composition that changes due to external processes (e.g., breathing, eating) and internal processes (e.g., metabolizing food).

Besides chemical transformations we'll also consider phase transformations. For example, solid ice and liquid water are composed of the same H_2O molecules so the transformation between phases may be written as,

$$H_2O\ (s) \rightleftharpoons H_2O\ (\ell)$$

where (s) and (ℓ) designate solid and liquid. In generic form this may be written as $\mathbb{A} \rightleftharpoons \mathbb{B}$. Phase transitions release or absorb energy, which is called the *latent heat* of the transformation. For example, for the transformation from solid to liquid the energy required is the latent heat of melting.

While living systems are usually considered to be under the purview of biology rather than physics there are interesting questions that may be addressed by thermodynamics, such as:

- What is life? That is, in terms of thermodynamics what distinguishes living systems from non-living systems?

- All animals consume food and produce waste matter; thermodynamically what's the difference between the two? And what about plants which "feed" on light by the process of photosynthesis? What did Schrödinger* mean when he said that living organisms feed on "negative entropy"?

- There are continuous processes by which needed materials (e.g., oxygen) are moved into and out of living cells. How is this accomplished and what is the role of the metabolism in sustaining these processes?

---

*E. Schrödinger. *What is Life - the Physical Aspect of the Living Cell*, Cambridge Univ. Press (1944).

Just as classical mechanics can explain the cat's ability to rotate in midair and land on its feet, thermodynamics can explain the basic metabolic processes that make the cat animate.

$$* * *$$

The illustrative examples in this chapter may seem mundane compared with topics such as black holes or quantum dots. Yet thermodynamics is universal so once you understand how it applies to the familiar you'll be well-equipped to tackle the exotic.

Chapter 2

Equations of State

The variables describing a thermodynamic system are not all independent. This chapter introduces some of their functional dependencies, such as the ideal gas law. To manipulate these functions we'll review some basic identities from multivariate calculus and then apply the results to model a simple thermodynamic process: the cooling of hot tea.

2.1   Equilibrium and Equations of State

Thermodynamic systems may be in or out of equilibrium.[*] To understand the difference let's return to our first example, the gas-filled balloon (see Section 1.1). For simplicity, we'll consider the case of a pure, inert gas (e.g., helium-filled balloon) and treat it as a homogeneous system. The next step is to select the state variables.

We can easily measure the volume, V, and mass, M, so these are convenient choices. Using mechanics (either classical or quantum) the energy of the gas may be defined. The system as a whole is taken to be at rest so we only consider internal energy, U. For our homogeneous system we will choose $\{U, V, M\}$ as the state variables, which means that it should be possible to express any thermodynamic quantity of interest as a function of these variables.

When the system is at *thermodynamic equilibrium* we can write the pressure as $p(U, V, M)$, which is called the *equation of state* for pressure. This function depends on the material properties, for example it's different for helium versus nitrogen. If the gas is a mixture of nitrogen and oxygen molecules then the equation of state has the form $p(U, V, M_{\mathrm{N}_2}, M_{\mathrm{O}_2})$. For a general mixture we write this as $p(U, V, \mathbf{M})$ where $\mathbf{M} = \{M_1, M_2, \ldots\}$. And the material

---

[*]By "equilibrium" we mean "thermodynamic equilibrium" as opposed to, say, mechanical equilibrium.

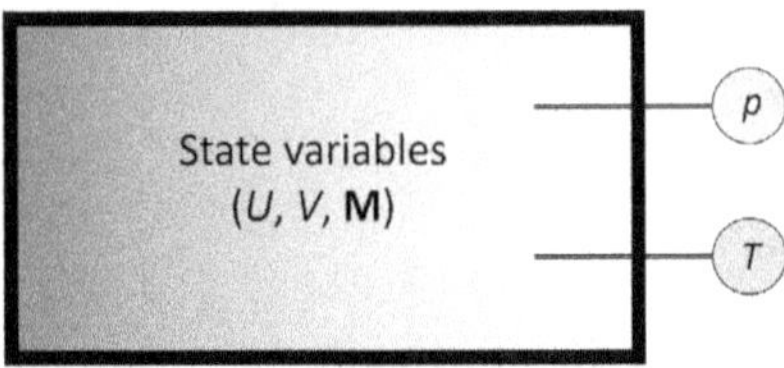

Figure 2.1: An isolated system with probes measuring pressure and temperature (i.e., a barometer and a thermometer). The system is initially in an inhomogeneous, non-equilibrium state.

in the system need not be a gas; liquid water and solid iron have their own equations of state for pressure.

At thermodynamic equilibrium we also have an equation of state that defines the temperature as $T(U, V, \mathbf{M})$. Given this function we can formulate $U(T, V, \mathbf{M})$ and then write the pressure in its more common form, $p(T, V, \mathbf{M})$ or $p(T, V, \mathbf{N})$ if you prefer moles instead of mass. As with pressure, the temperature equation of state depends on the material comprising the system as well as its phase (i.e., liquid water is different from solid ice). For open and reactive systems there's yet another equation of state called the chemical potential, $\mu_k(U, V, \mathbf{M})$, which is discussed in Chapter 9.

When the system is *not* at equilibrium the variables U, V, and $\mathbf{M}$ are still well-defined but the thermodynamic pressure and temperature are not defined, at least not if we want to model the balloon as a homogeneous system. Consider the isolated system illustrated in Figure 2.1 that starts in an inhomogeneous, non-equilibrium state. The barometer and thermometer measurements will vary with time despite the fact that in an isolated, inert system U, V, and $\mathbf{M}$ are constant. The pressure and temperature measurements agree with the equations of state $p(U, V, \mathbf{M})$ and $T(U, V, \mathbf{M})$ only after the system relaxes to thermodynamic equilibrium. As Callen[†] puts it, "a system is in an equilibrium state if its properties are consistently described by thermodynamic theory!" That said, later in this chapter you'll see how modern thermodynamics treats non-equilibrium scenarios.

Equations of State for Gases

An equation of state that you're already familiar with is the *ideal gas law*. Instead of writing it as $p(U, V, M)$ the traditional form is

$$p(T, V, N) = \frac{NRT}{V} \qquad \text{(Pure Ideal Gas)}$$

[†]H.B. Callen, *Thermodynamics and an Introduction to Thermostatistics*, Wiley (1985).

where $R = k_B N_A$ is the gas constant (see Table 1.1). For a mixture $p = \sum_k N_k RT/V$. Similarly, instead of writing the temperature equation of state, $T(U, V, M)$, for a pure ideal gas one writes

$$U(T, V, M) = c_V M(T - T_0) + M u_0$$

or

$$U(T, V, N) = \hat{c}_V N(T - T_0) + N \hat{u}_0$$

where c_V and $\hat{c}_V$ are the specific and molar heat capacities at constant volume.[‡] To keep things simple we'll take $T_0 = u_0 = \hat{u}_0 = 0$. From the above, the equations of state for the pure ideal gas are

$$p(U, V, M) = \frac{k_B U}{c_V m V} \qquad \text{and} \qquad T(U, V, M) = \frac{U}{c_V M}.$$

A simple, but often accurate, approximation for the molar heat capacity is

$$\hat{c}_V = \tfrac{1}{2} f_d R$$

where f_d is the number of classical degrees of freedom for the gas molecules. For monatomic gases (e.g., helium, krypton) $f_d = 3$ while for diatomic gases (e.g., nitrogen, oxygen) $f_d = 5$. Finally, for mixtures of ideal gases we can simply sum the contributions for each species, for example $U = \sum_k c_{V,k} M_k T$ where the sum is over species.

Example 2.1: On a cold winter morning you turn on an electric heater and raise the temperature in your room. The room is not air-tight so it remains at atmospheric pressure as the temperature increases. Show that the internal energy of the air in the room does *not* change.
Solution: Given that $U = \hat{c}_V N T$ you might think that the internal energy would increase with temperature. However the amount of air in the room is not constant (i.e., the room is an open system). From the equation of state for pressure,

$$U = \frac{\hat{c}_V}{R} \, pV$$

so internal energy is constant for fixed pressure and volume. You might object that air is not a pure substance but you can show that the same result applies to gas mixtures.

The ideal gas law is only accurate for dilute gases; a better model is the *van der Waals equation*

$$p = \frac{NRT}{V - Nb} - a \left(\frac{N}{V}\right)^2$$

[‡] Heat capacity is formally defined in Chapter 4.

for $V > Nb$. The coefficients a and b model the intermolecular forces of attraction and repulsion, respectively. The internal energy for a van der Waals gas is

$$U = \hat{c}_V N T - \frac{aN^2}{V}.$$

Notice that U is a function of both temperature and volume for a van der Waals gas.

Example 2.2:   Using the van der Waals and the ideal gas equations of state calculate the pressure for 20 moles of molecular nitrogen gas in a 10 liter vessel at a temperature of 300 Kelvin. The van der Waals coefficients for nitrogen are $a = 1.35$ L^2 atm/mol^2 and $b = 0.0387$ L/mol.
Solution: From the van der Waals equation $p = 48$ atmospheres and from the ideal gas law, $p = 49$ atmospheres.

2.2   Partial Derivatives

Thermodynamic coefficients

The relations among thermodynamic variables are often in the form of thermodynamic coefficients. One example is c_V, the specific heat capacity. Another is the *coefficient of thermal expansion,*

$$\alpha = \frac{1}{V} \left(\frac{d}{dT} \right) V(T, p, \mathbf{M}) \qquad \text{(fixed } p, \mathbf{M}).$$

This coefficient is the fractional change in volume due to a change in temperature when pressure and mass are fixed. For example, for iron $\alpha \approx 3.5 \times 10^{-5}$ K^{-1} so a 10 degree increase in temperature results in about a 0.035% increase in volume.

We will write this as the partial derivative

$$\alpha = \frac{1}{V} \left(\frac{\partial V}{\partial T} \right)_{p, \mathbf{M}}.$$

The subscript notation is to remind us to hold certain quantities fixed when we take the derivative (e.g., hold p and $\mathbf{M}$ fixed when taking the derivative with respect to temperature). This notation is useful in thermodynamics because for the same quantity we may pick different state variables depending on the problem. For example, sometimes we'll have $U(T, V, \mathbf{M})$ and other times we'll be working with $U(T, p, \mathbf{M})$. It's important to realize that

$$\left(\frac{\partial U}{\partial T} \right)_{V, \mathbf{M}} \neq \left(\frac{\partial U}{\partial T} \right)_{p, \mathbf{M}}$$

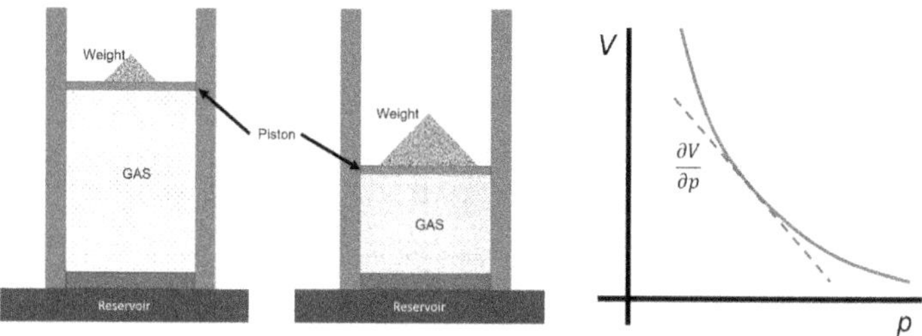

Figure 2.2: Illustration of isothermal compression; volume decreases as pressure increases and the slope is used to find the isothermal compressibility, κ_T.

and the subscripts remind us what we're using as the independent variables.

Another useful coefficient is the *isothermal compressibility*,

$$\kappa_T = -\frac{1}{V}\left(\frac{\partial V}{\partial p}\right)_{T,\mathrm{M}},$$

which is the fractional change in volume due to a change in pressure when temperature and mass are fixed (see Fig. 2.2). For example, for iron $\kappa_T \approx 5.9 \times 10^{-12}$ Pa^{-1} so increasing the pressure by 10^8 pascals (~ 1000 atmospheres) results in about a 0.059% decrease in volume.

Example 2.3: Find α and κ_T for a pure ideal gas.

Solution: From the ideal gas law,

$$\alpha = \frac{1}{V}\left(\frac{\partial V}{\partial T}\right)_{p,\mathrm{M}} = \frac{1}{V}\left(\frac{\partial}{\partial T}\right)_{p,\mathrm{M}}\frac{NRT}{p} = \frac{NR}{pV} = \frac{1}{T}.$$

Notice that since $N = M/(m\mathrm{N_A})$ holding N fixed is the same as holding M fixed. Similarly,

$$\kappa_T = -\frac{1}{V}\left(\frac{\partial V}{\partial p}\right)_{T,\mathrm{M}} = -\frac{1}{V}\left(\frac{\partial}{\partial p}\right)_{T,\mathrm{M}}\frac{NRT}{p} = \frac{NRT}{p^2 V} = \frac{1}{p}.$$

By the way, the same results are obtained for gas mixtures.

Manipulating Partial Derivatives

In thermodynamics calculations we often need to manipulate partial derivatives using these mathematical identities:

$$\left(\frac{\partial x}{\partial y}\right)_w \left(\frac{\partial y}{\partial z}\right)_w = \left(\frac{\partial x}{\partial z}\right)_w \qquad \text{(D1)};$$

$$\left(\frac{\partial x}{\partial y}\right)_z = \frac{1}{\left(\frac{\partial y}{\partial x}\right)_z} \qquad \text{(D2)};$$

$$\left(\frac{\partial x}{\partial y}\right)_z \left(\frac{\partial y}{\partial z}\right)_x \left(\frac{\partial z}{\partial x}\right)_y = -1 \qquad \text{(D3)}.$$

Identity (D3) is known as the cyclic chain rule or triple product rule; sometimes it is written as

$$\left(\frac{\partial x}{\partial z}\right)_y = -\left(\frac{\partial x}{\partial y}\right)_z \left(\frac{\partial y}{\partial z}\right)_x \qquad \text{(D3$'$)}.$$

With all of these identities we can add extra constrained variables to all the derivatives, for example,

$$\left(\frac{\partial x}{\partial z}\right)_{y,w} = -\left(\frac{\partial x}{\partial y}\right)_{z,w} \left(\frac{\partial y}{\partial z}\right)_{x,w}$$

for the cyclic chain rule.

For the function $f(x,y)$ the exact differential df is defined as

$$df(x,y) = \left(\frac{\partial f}{\partial x}\right)_y dx + \left(\frac{\partial f}{\partial y}\right)_x dy$$

and

$$\left(\frac{\partial}{\partial x}\right)_y \left(\frac{\partial f}{\partial y}\right)_x = \left(\frac{\partial}{\partial y}\right)_x \left(\frac{\partial f}{\partial x}\right)_y = \frac{\partial^2 f}{\partial x \partial y} \qquad \text{(D4)}.$$

We also have,

$$\left(\frac{\partial f}{\partial x}\right)_z = \left(\frac{\partial f}{\partial x}\right)_y + \left(\frac{\partial f}{\partial y}\right)_x \left(\frac{\partial y}{\partial x}\right)_z \qquad \text{(D5)},$$

which is a variant of the two-dimensional chain rule.

Example 2.4: Express the rate at which pressure changes with temperature when V and M are fixed in terms of α and κ_T.

Solution: The desired quantity is,

$$
\begin{aligned}
\left(\frac{\partial p}{\partial T}\right)_{V,\mathbf{M}} &= -\left(\frac{\partial p}{\partial V}\right)_{T,\mathbf{M}}\left(\frac{\partial V}{\partial T}\right)_{p,\mathbf{M}} && \text{(Using D3$'$)} \\
&= -\left(\frac{\partial V}{\partial T}\right)_{p,\mathbf{M}}\bigg/\left(\frac{\partial V}{\partial p}\right)_{T,\mathbf{M}} && \text{(Using D2)} \\
&= -(V\alpha)/(-V\kappa_T) = \alpha/\kappa_T .
\end{aligned}
$$

Equations of State for Simple Liquids and Solids

The ideal gas law is a good approximation for a dilute gas, such as the air in your room. But what about a glass of water or an iron nail? We know from experience that the volume of a liquid or a solid varies only slightly with pressure and temperature. Let's formulate approximate equations of state for modelling simple liquids and solids.

For a pure system, the equation of state for pressure is $p(U, V, M)$ and using the equation of state for temperature we can write pressure as $p(T, V, M)$. Let's rearrange the variables and write $V(T, p, M)$. Recall from calculus that for a function of three variables, $f(x, y, z)$, the differential of the function is

$$
df = \left(\frac{\partial f}{\partial x}\right)_{y,z} dx + \left(\frac{\partial f}{\partial y}\right)_{x,z} dy + \left(\frac{\partial f}{\partial z}\right)_{x,y} dz.
$$

Given $V(T, p, M)$ we can write the differential

$$
dV = \left(\frac{\partial V}{\partial T}\right)_{p,M} dT + \left(\frac{\partial V}{\partial p}\right)_{T,M} dp + \left(\frac{\partial V}{\partial M}\right)_{T,p} dM.
$$

Let's take a closed system (so $dM = 0$) and use the definitions of α and κ_T to write this as

$$
dV = \alpha V dT - \kappa_T V dp.
$$

This result also holds for mixtures as long as $\mathbf{M}$ is fixed.

For simple liquids and solids we'll assume α and κ_T are constant. As an exercise you can show that by integration

$$
V(T, p, M) = V_0 \exp\{\alpha(T - T_0) - \kappa_T(p - p_0)\}
$$

where T_0 and p_0 are the temperature and pressure at a convenient reference state and $V_0 = V(T_0, p_0, M)$. For liquids and solids both α and κ_T are small, which justifies Taylor expanding the exponential and writing

$$
V(T, p, M) = V_0 \left(1 + \alpha(T - T_0) - \kappa_T(p - p_0)\right)
$$

after dropping higher order terms. From this

$$p(T, V, M) = p_0 + \frac{\alpha}{\kappa_T}(T - T_0) + \frac{1}{\kappa_T}(1 - V/V_0)$$

although, in practice, the expression for $V(T, p, M)$ is more useful.

Example 2.5: An iron cup is filled to the brim with 100 cm^3 of water; both the cup and the water are at 20 °C. How much water (if any) spills out of the cup when their temperature increases to 80 °C? Take $\alpha = 3.5 \times 10^{-5}$ K^{-1} for iron and $\alpha = 2.1 \times 10^{-4}$ K^{-1} for water; assume constant atmospheric pressure.
Solution: The increase in volume of the water is

$$V_{\text{water}} = V_0(1 + \alpha(T - T_0)) = (100 \text{ cm}^3)(1 + (2.1 \times 10^{-4} \text{ K}^{-1})(60 \text{ K})) = 101.26 \text{ cm}^3.$$

The volume of the interior of the iron cup also increases to $V_{\text{cup}} = 100.21$ cm^3 so about one cubic centimeter of water is spilled.

The relation between temperature and internal energy for simple liquids and solids can be written as

$$U(T, p, \mathbf{M}) \approx U_0 + c_p M(T - T_0) + \mathcal{C}M(p - p_0).$$

The coefficient $\mathcal{C}$, which depends on α and κ_T, is small so it's often neglected. In general the specific heat capacity is itself a function of temperature, pressure, and composition (e.g., pure water versus salt water). That said, for simple liquids and solids we can approximate it as constant. The Dulong-Petit law, $\hat{c}_p = 3R$, is often accurate for elementary solids such as iron.

Finally, these equations of state for simple liquids and solids are accurate for standard conditions (i.e., atmospheric pressure and room temperature). For exotic conditions, such as near absolute zero, there are alternative equations of state (e.g., Debye model for heat capacity).

2.3 Thermodynamic Processes

Section 1.2 describes the tea cup example (see Figures 1.2 and 2.3) which we model as a pair of homogeneous closed subsystems (the liquid tea and the solid cup). The subsystems (tea and cup) may exchange energy but we'll neglect any interactions with the environment, modeling the total system as isolated.

The subsystems start at thermodynamic equilibrium but at different temperatures, specifically hot tea in a cold cup. In this case $U_{\text{tea}}(t)$ and $U_{\text{cup}}(t)$ are functions of time and from experience we expect energy will be transferred from the tea into the cup until the total system reaches equilibrium

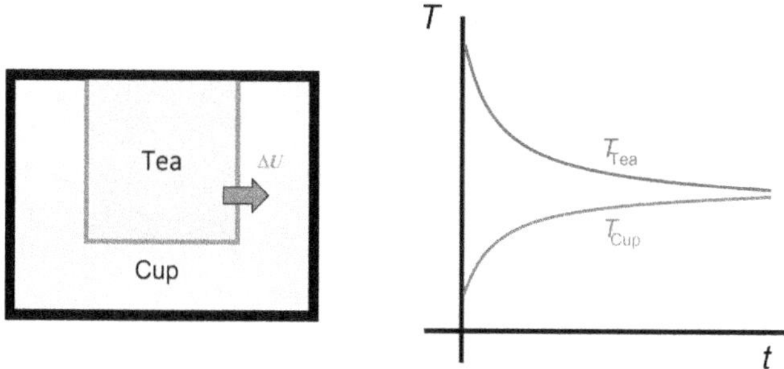

Figure 2.3: An isolated, heterogeneous system consisting of a pair of closed subsystems, hot tea and a cold cup. Over time the temperature of the tea decreases and that of the cup increases.

with $T_{\text{tea}} = T_{\text{cup}}$. These changes and how they occur are generically referred to as a *thermodynamic process*.

During this process we can always define and measure U_{tea}. However we can only define T_{tea} when the tea is at thermodynamic equilibrium since the equation of state $T(U, V, \mathbf{M})$ is only valid at equilibrium. If the heat conduction is very slow then we may assume that the tea is always in thermodynamic equilibrium. In that case we can write

$$T_{\text{tea}}(t) = T_{\text{tea}}(U_{\text{tea}}(t), V, \mathbf{M}).$$

The cooling of the tea occurs by a *quasi-static process*; the "quasi-" means that the temperature is changing but the tea is always close to an equilibrium state.

Example 2.6: The tea and cup have initial temperatures of $T_{\text{tea}} = 90\ °C$ and $T_{\text{cup}} = 20\ °C$. What is the temperature of the cup when the tea has cooled to $T'_{\text{tea}} = 85\ °C$. Assume quasi-static processes; the specific heat capacities are $c_{\text{tea}} = 1.0\ \text{cal}/(\text{K g})$, $c_{\text{cup}} = 0.2\ \text{cal}/(\text{K g})$ and the masses are $M_{\text{tea}} = 100\ \text{g}$ and $M_{\text{cup}} = 80\ \text{g}$.

Solution: The total internal energy is initially

$$U_\Sigma = U_{\text{tea}} + U_{\text{cup}} - c_{\text{tea}} M_{\text{tea}} T_{\text{tea}} + c_{\text{cup}} M_{\text{cup}} T_{\text{cup}}.$$

The system is isolated so when $U'_{\text{tea}} = c_{\text{tea}} M_{\text{tea}} T'_{\text{tea}}$ then $U'_{\text{cup}} = U_\Sigma - U'_{\text{tea}} = U_{\text{cup}} + (U_{\text{tea}} - U'_{\text{tea}})$ so

$$T'_{\text{cup}} - \frac{U'_{\text{cup}}}{c_{\text{cup}} M_{\text{cup}}} = T_{\text{cup}} + \frac{c_{\text{tea}} M_{\text{tea}}}{c_{\text{cup}} M_{\text{cup}}}(T_{\text{tea}} - T'_{\text{tea}}) = 51\ °C.$$

Finally, notice that we omitted the constant term U_0 in the internal energy for the tea and the cup; you can check that it cancels out in the calculation of temperature.

For the total system (tea and cup) we can always measure the total internal energy, $U_\Sigma = U_{\text{tea}} + U_{\text{cup}}$, and similarly the total volume and mass. However as long as $T_{\text{tea}} \neq T_{\text{cup}}$ we may *not* define temperature for the total system as $T_\Sigma(U_\Sigma, V_\Sigma, \mathbf{M}_\Sigma)$. The reason is that the total system is heterogeneous so even if the thermodynamic process occurs slowly it is out of equilibrium if the two subsystems are at different temperatures. So while each subsystem is in *local equilibrium*, the total system is not in equilibrium.

The total system starts and remains out of equilibrium until, eventually, it reaches equilibrium with $T_{\text{tea}} = T_{\text{cup}} = T_\Sigma(U_\Sigma, V_\Sigma, \mathbf{M}_\Sigma)$. This means that we have a *non-equilibrium process*, which is the opposite of a quasi-static process. Notice that we can have quasi-static processes for the tea and for the cup yet have a non-equilibrium process for the total system. The significance of this observation will become clear when we calculate entropy production in Chapter 7.

Example 2.7:   Using the values from the previous example, find the final temperature for the tea and cup, T_Σ, when the total system reaches thermodynamic equilibrium.
Solution: The system is isolated so the initial internal energy equals the final internal energy at equilibrium

$$U_\Sigma^{\text{eq}} = c_{\text{tea}} M_{\text{tea}} T_\Sigma + c_{\text{cup}} M_{\text{cup}} T_\Sigma = c_\Sigma M_\Sigma T_\Sigma$$

where $M_\Sigma = M_{\text{tea}} + M_{\text{cup}}$. This gives

$$T_\Sigma = \frac{c_{\text{tea}} M_{\text{tea}} T_{\text{tea}} + c_{\text{cup}} M_{\text{cup}} T_{\text{cup}}}{c_{\text{tea}} M_{\text{tea}} + c_{\text{cup}} M_{\text{cup}}.} = 80 \ {}^\circ\text{C}.$$

This result only assumes equilibrium for the initial states of the tea and cup so it applies to both quasi-static and non-equilibrium processes.

---

* * *

At this point it's clear that internal energy is a key element in thermodynamics. The next chapter introduces the First Law of Thermodynamics, which describes the processes by which the energy of a system can change due to heat and work.

Chapter 3

Work and the First Law

The First Law of Thermodynamics is essentially just the principle of conservation of energy for thermodynamic systems. Energy may be exchanged between two systems but according to the First Law the amount of energy gained by one system must exactly equal the energy lost by the other system. The tricky part is that energy can take various forms so we have to carefully "balance the books."

3.1   First Law of Thermodynamics

Internal Energy

Consider a homogeneous system with a given equation of state for temperature, $T(U, V, \mathbf{M})$. From this we can write the internal energy as $U(T, V, \mathbf{M})$. Using temperature as a state variable implies that we take the system to be at thermodynamic equilibrium. For simplicity we'll limit ourselves to closed, inert systems so $\mathbf{M}$ is fixed and we'll simply write $U(T, V)$.

In classical mechanics we define gravitational potential energy relative to a convenient reference height (e.g., $y = 0$ is the floor). Similarly, it is useful to define *reference states* in thermodynamics. For example, the reference state could be $U_0 = U(T_0, V_0)$ for $T_0 = 273$ K, $p_0 = 1.0$ atm, $M_0 = 1.0$ kg with V_0 given by the equation of state. As with potential energy we're often only interested in changes, such as $\Delta U_{AB} = U_B - U_A$, rather than U_A and U_B individually, and this change is independent of the choice of reference state.

Now suppose we start a system at an equilibrium state (T_A, V_A) then change the temperature and volume to reach a final equilibrium state (T_B, V_B). Since internal energy is a state variable

$$\Delta U_{AB} = U_B - U_A = U(T_B, V_B) - U(T_A, V_A)$$

and this is true regardless of how we go from state A to state B.

To define a quasi-static process that takes the system from state A to B we can use the functions

$$T = \mathfrak{T}(t) \quad \text{where} \quad \mathfrak{T}(t_A) = T_A, \ \mathfrak{T}(t_B) = T_B$$
$$V = \mathfrak{V}(t) \quad \text{where} \quad \mathfrak{V}(t_A) = V_A, \ \mathfrak{V}(t_B) = V_B.$$

The parameter t goes from $t = t_A$ to $t = t_B$ as the system goes from state A to state B along the path defined by $\mathfrak{T}(t)$ and $\mathfrak{V}(t)$. This is a quasi-static process since at every point along the path the system is at thermodynamic equilibrium, which allows us to define the value of temperature. The change in the internal energy is given by the path integral

$$\Delta U_{AB} = \int_A^B dU = \int_{t_A}^{t_B} \left\{ \frac{d}{dt} U(\mathfrak{T}(t), \mathfrak{V}(t)) \right\} dt.$$

The important point here is that the integral is the *same* no matter what we choose for the functions $\mathfrak{T}(t)$ and $\mathfrak{V}(t)$.

Example 3.1: Find ΔU_{AB} for a pure ideal gas ($U(T, V) = c_V MT$) for the path $\mathfrak{T}(t) = T_A + t^3 \Delta T$ where $t_A = 0$, $t_B = 1$, and $\Delta T = T_B - T_A$.
Solution: For an ideal gas the internal energy does not depend on volume so

$$\frac{d}{dt} U(\mathfrak{T}(t), \mathfrak{V}(t)) = \frac{d}{dt} c_V M \mathfrak{T}(t) = 3c_V M \Delta T t^2$$

for arbitrary $\mathfrak{V}(t)$. The integral gives

$$\Delta U_{AB} = 3c_V M \Delta T \int_0^1 t^2 dt = c_V M \Delta T,$$

which is the same as $U(T_B, V_B) - U(T_A, V_A)$.

In fact ΔU_{AB} is the same even if we go from state A to B by a non-equilibrium process (i.e., not quasi-static). For such a process the temperature of the system is undefined except at the initial and final states. Figure 3.1 illustrates two different processes that go from state A to state B, one is a quasi-static process that has a path in the $T - V$ diagram and the other is a non-equilibrium process, which does not.

Heat and Work

To go further we need to specify the mechanisms by which the energy of a system changes in going from state A to state B. One mechanism is heat, which is usually thought of as a transfer of energy occurring between objects at different temperatures. Later we'll develop a precise definition of heat once we've defined entropy.

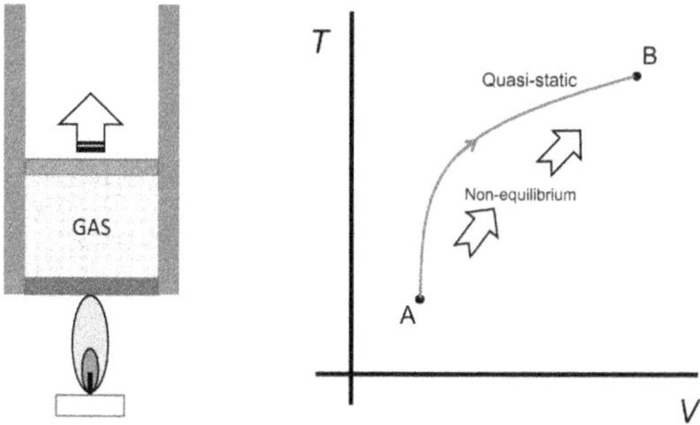

Figure 3.1: Gas being heated and expanding in a closed system. The $T-V$ diagram shows two processes: a quasi-static path (curve) and a non-equilibrium process (arrows). The change of internal energy, ΔU_{AB}, is the same for any process going from state (T_A, V_A) to (T_B, V_B).

Besides thermal heat a system may gain or lose energy by mechanical work. For example, when a force compresses an elastic spring this force does work on the system (the spring) and increases its energy in the form of elastic potential energy.

The *First Law of Thermodynamics* says that the change in internal energy for a system is

$$\Delta U_{AB} = Q_{AB} + W_{AB} \qquad \text{or} \qquad dU = dQ + dW$$

where Q is the energy added by heat and W is the energy added by work.

Note that our definition of W is the mechanical work done *on* the system so if $W > 0$ then energy is added to the system via work. In some situations (e.g., engines) it's convenient to use $\mathcal{W}$, the work done *by* the system; by definition $\mathcal{W} = -W$. Note that some textbooks prefer to use $\mathcal{W}$ so they'll write $\Delta U_{AB} = Q_{AB} - \mathcal{W}_{AB}$ as the First Law.

Again, consider going from an initial equilibrium state (T_A, V_A) to a final equilibrium state (T_B, V_B) by a process defined by $\mathfrak{T}(t)$ and $\mathfrak{V}(t)$. The tricky and confusing thing about heat and work is that the path integrals

$$Q_{AB} = \int_A^B dQ \qquad \text{and} \qquad W_{AB} = \int_A^B dW$$

do depend on the choice of the functions $\mathfrak{T}(t)$, and $\mathfrak{V}(t)$, that is, on the path between states A and B. For example, suppose that $U_A = 2$ Joules and $U_B = 5$ Joules so $\Delta U_{AB} = 3$ Joules. For one path between these states we might have $Q_{AB} = -1$ Joules and $W_{AB} = 4$ Joules while on a different path we could have $Q_{AB} = W_{AB} = 1.5$ Joules.*

---

*Because Q_{AB} and W_{AB} depend on the path some textbooks use the notation $đQ$ and $đW$ to remind the reader that they are not exact differentials.

As a side note, the First Law also applies to open, reactive systems. When mass and composition are allowed to vary we write

$$\Delta U_{AB} = U(T_B, V_B, \mathbf{M_B}) - U(T_A, V_A, \mathbf{M_A})$$

and ΔU_{AB} is the same regardless of the process that takes us from state A to state B. We still write the First Law as $\Delta U_{AB} = Q_{AB} + W_{AB}$ with heat and work depending on the process.

3.2   Compression Work

Keeping in mind that the path matters, let's see how to compute W. The mechanical work done by an external force, $\mathbf{F_e}$, that acts over a distance, $d\mathbf{x}$, is $dW = \mathbf{F_e} \cdot d\mathbf{x}$. Since we are using volume as a state variable we'll use pressure instead of force; multiply and divide by area, A, and you get

$$dW = - \left(\frac{F_e}{A}\right)(A dx) = -p_e dV$$

where the minus sign comes from the fact that $dW > 0$ when $dV < 0$ (i.e., compress the system). This type of mechanical work is called *compression work*.

For a quasi-static process the pressure due to the external force, p_e, equals the pressure in the system, p, given by the equation of state. In that case we have

$$dW = -pdV \qquad \text{and} \qquad W_{AB} = - \int_A^B pdV.$$

On a quasi-static path we can use the equation of state for pressure, $p(U, V, \mathbf{M})$, to evaluate the integral, as shown in the next example.

Example 3.2: Consider a closed system with N moles of a pure inert gas held at constant temperature T^{res} by a thermal reservoir. Find the work done on the system in going from volume V_A to V_B by a quasi-static process (see Figure 3.3). **Solution:** To find the work done on the system we evaluate

$$W_{AB} = - \int_{V_A}^{V_B} p(V)\, dV.$$

The path from state A to state B is uniquely defined by the fact that the temperature is fixed at T^{res} (see Figure 3.2). We use the equation of state for an ideal gas, expressed in its more familiar form, $pV = NRT$. From the equation of state $p(V) = NRT^{res}/V$ so

$$W_{AB} = -NRT^{res} \int_{V_A}^{V_B} \frac{dV}{V} = -NRT^{res}(\ln V_B - \ln V_A) = NRT^{res} \ln(V_A/V_B).$$

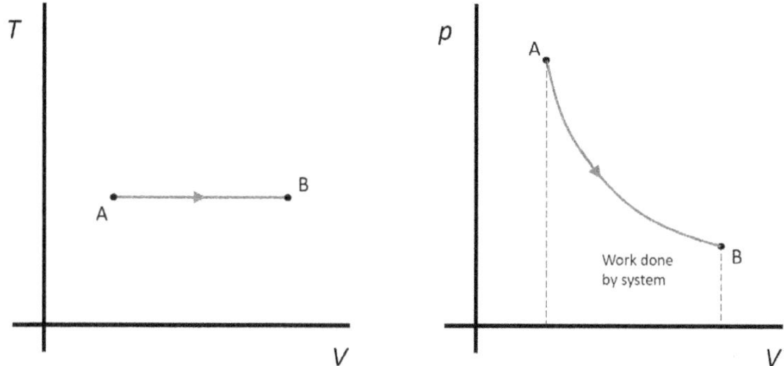

Figure 3.2: A quasi-static, isothermal expansion from volume V_A to V_B on the: $T - V$ diagram (left); $p - V$ diagram (right).

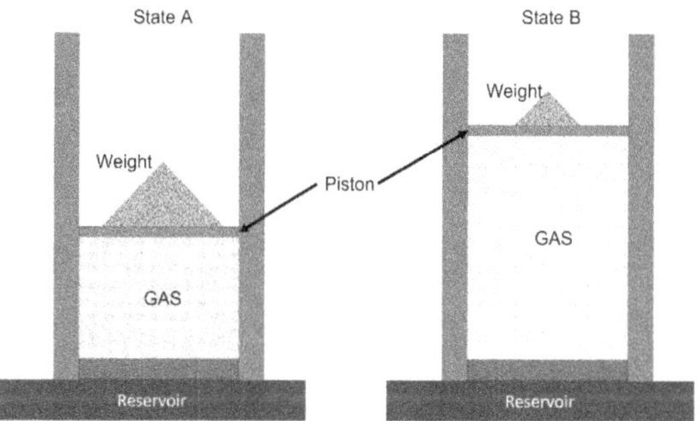

Figure 3.3: An isothermal expansion for an ideal gas held at fixed temperature by a thermal reservoir. Weight is removed from the piston to reduce the pressure on the system. The system does work in lifting the weight remaining on the piston.

For example, if $V_A < V_B$ then the system expands so $W_{AB} < 0$, that is, a negative amount of work is done on the system. This means that $\mathcal{W}_{AB} = -W_{AB} > 0$ so a positive amount of work is done *by* the system. Figure 3.2 illustrates that $\mathcal{W}_{AB} = \int_A^B p\,dV$ equals the area under the path curve in the $p - V$ diagram. Note that on a different path we would have to know how temperature varied with volume (i.e., be given $T(V)$ along the path) and use $p(V) = NRT(V)/V$ as the integrand.

Example 3.3: For the system in the previous example find the heat added in going from state A to state B. Assume that the internal energy for the ideal gas is $U = c_V M T$ where the specific heat capacity is constant.

Solution: We found that the work done on the system is

$$W_{AB} = NRT^{\text{res}} \ln(V_A/V_B)$$

Since it's a closed, inert system (M fixed) and the process is isothermal ($T = T^{\text{res}}$ fixed) then $\Delta U_{AB} = 0$ given that $U = c_V M T$ for this system.

By the First Law, $\Delta U_{AB} = W_{AB} + Q_{AB}$ so

$$Q_{AB} = -W_{AB} = -NRT^{\text{res}} \ln(V_A/V_B).$$

For example, if $V_A < V_B$ then the system expands so $Q_{AB} > 0$, which means that energy enters the system in the form of heat. As work is done by the system, heat enters the system to keep the temperature constant.

Finally, note that in general $Q_{AB} \neq -W_{AB}$; this example is a special case since $\Delta U_{AB} = 0$ for an ideal gas in an isothermal process. In the next chapter we'll see how to obtain Q_{AB} in other scenarios.

Example 3.4: Find the work done in an isothermal expansion ($T = T^{\text{res}}$) of N moles from an initial volume V_A to a final volume V_B using the equation of state for a van der Waals gas (see Section 2.1).

Solution: Using the van der Waals equation for the pressure

$$W_{AB} = -\int_{V_A}^{V_B} p\,dV = -NRT^{\text{res}} \int_{V_A}^{V_B} \frac{1}{V - Nb}\,dV + aN^2 \int_{V_A}^{V_B} \frac{1}{V^2}\,dV.$$

Evaluating these integrals gives

$$W_{AB} = -NRT^{\text{res}} \ln\left(\frac{V_B - Nb}{V_A - Nb}\right) + aN^2 \left(\frac{1}{V_A} - \frac{1}{V_B}\right).$$

Notice that this result matches the earlier result for an ideal gas when $a = 0$, $b = 0$. By the way, for the van der Waals gas the internal energy depends on volume so $Q_{AB} \neq -W_{AB}$.

3.3 Mechanical Work*

The previous section introduced mechanical work with the example of compression work, $dW = -p\,dV$, for systems using pressure and volume as state variables. Yet in some cases there are other types of mechanical or electromagnetic forces doing mechanical work on a system. For example, for a wire under tension the work done on the wire is the applied force times the change in length.

Generalized Force, $\mathfrak{F}$	Generalized Displacement, $\mathfrak{X}$
Pressure	Volume
Surface tension	Surface area
Tension	Length
Torque	Angle
Stress	Strain
Electro-motive Force	Electric charge
Electric field	Electric dipole moment
Magnetic field	Magnetic dipole moment

Table 3.1: List of generalized forces and displacements.

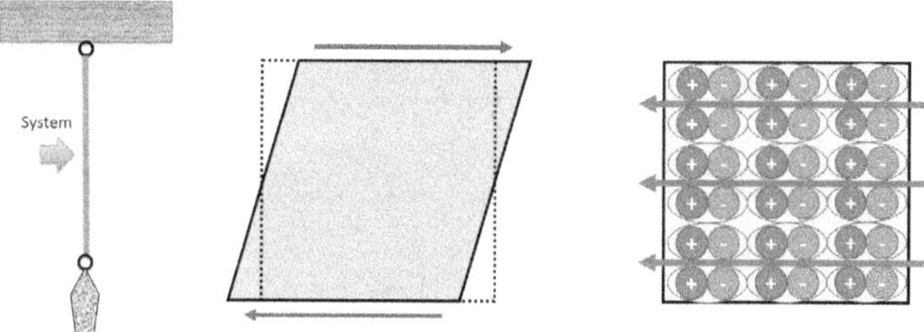

Figure 3.4: Examples of work done on a system by: (left) tension force stretching a wire; (middle) shearing stress creating strain; (right) electric field creating polarization.

Introducing the generalized force, $\mathfrak{F}$, and generalized displacement, $\mathfrak{X}$, we write the generalized form of mechanical work as

$$dW = \mathfrak{F}\,d\mathfrak{X}$$

with $\mathfrak{X}$ as an extensive state variable and $\mathfrak{F}$ as an intensive state variable. Figure 3.4 illustrates and Table 3.1 lists some common scenarios. Note that for compression work the generalized displacement is $\mathfrak{X} = V$ and the generalized force is $\mathfrak{F} = -p$ with the minus sign appearing since, at constant pressure, work is done on the system when the volume *decreases*.

Example 3.5: For a wire (see Fig. 3.4) the total length, ℓ, is used as a state variable instead of volume. For quasi-static processes the applied force is balanced by the tension force, f. By Hooke's law

$$f = k_0[\ell - \ell_0(M)]$$

where the rest length is $\ell_0(M)$ and k_0 is the elastic constant. Find ΔU_{AB} when the wire is stretched from the rest length $\ell_{\mathrm{A}} = \ell_0$ to ℓ_{B} by a quasi-static adiabatic process.

Solution: For this system $\mathfrak{F} = f$ and $\mathfrak{X} = \ell$. The process is adiabatic ($dQ = 0$) so

$$\Delta U_{\mathrm{AB}} = \int_{\mathrm{A}}^{\mathrm{B}} dW = \int_{\ell_0}^{\ell_{\mathrm{B}}} f(\ell)d\ell = \int_{\ell_0}^{\ell_{\mathrm{B}}} k_0[\ell - \ell_0]d\ell = \frac{1}{2}k_0[\ell_{\mathrm{B}} - \ell_0]^2,$$

which is the elastic potential energy.

For some systems we may be interested in more than one generalized force. For example, work could be done on a gas both by pressure changing the volume and by an applied electric field changing the dipole moment (i.e., total polarization in a dielectric). The most general form is

$$dW = \sum_{j=1}^{N_{\mathrm{w}}} \mathfrak{F}_j d\mathfrak{X}_j$$

for N_{w} types of mechanical work. This can be written in a compact "vector" form as

$$dW = \mathfrak{F} \cdot d\mathfrak{X}$$

where $\mathfrak{F} = \{\mathfrak{F}_1, \ldots, \mathfrak{F}_{N_{\mathrm{w}}}\}$, $\mathfrak{X} = \{\mathfrak{X}_1, \ldots, \mathfrak{X}_{N_{\mathrm{w}}}\}$, and the dot product indicates a sum over the elements.

Having said all of this, due to tradition and for notational convenience we will rarely use this general form. For mechanical work we'll mostly limit ourselves to compression work, $dW = -pdV$. Just remember that, if needed, V can be replaced with $\mathfrak{X}$, in which case $dW = -pdV$ is replaced by $dW = \mathfrak{F} \cdot d\mathfrak{X}$.

Example 3.6:   An isolated, heterogeneous system consists of a spherical droplet of liquid (radius R) suspended in vapor (see Figure 3.5). The work done on each of these subsystems is

$$dW_\ell = -p_\ell dV_\ell + \gamma_{\mathrm{st}}dA_\ell \qquad \text{and} \qquad dW_v = -p_v dV_v$$

where V_ℓ and A_ℓ are the volume and surface area of the liquid droplet; γ_{st} is the surface tension of the liquid. Show that if $dQ_\ell = dQ_v = 0$ then $R = 2\gamma_{\mathrm{st}}/\Delta p$ where $\Delta p = p_\ell - p_v$ is the *Laplace pressure*.

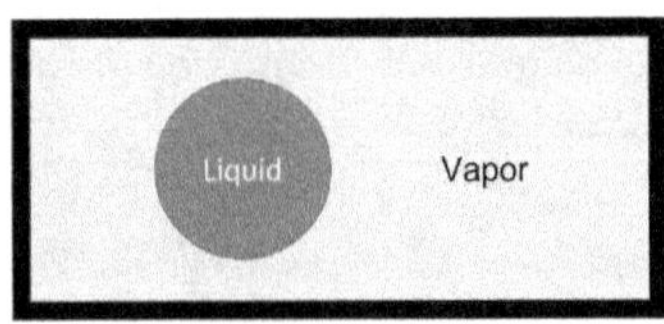

Figure 3.5: Spherical liquid droplet (radius R) suspended in vapor.

Solution: For an isolated system $dU = dU_\ell + dU_v = 0$ so $dW_\ell + dW_v = 0$. Since $V_\ell = \frac{4}{3}\pi R^3$ and $A_\ell = 4\pi R^2$ then $dV_\ell = 4\pi R^2 dR$ and $dA_\ell = 8\pi R dR$. Using these results along with the fact that the total volume is constant so $dV_v = -dV_\ell$ gives

$$\left[-p_\ell(4\pi R^2 dR) + \gamma_{st}(8\pi R dR)\right] + \left[-p_v(-4\pi R^2 dR)\right] = 0.$$

Collecting terms gives $\{(p_v - p_\ell)R + 2\gamma_{st}\}dR = 0$, which gives the desired result.

$$* * *$$

The First Law points out the duality of energy changes by thermodynamic processes. This chapter described work so the next chapter treats its complement, heat.

Chapter 4

Heat and the First Law

The First Law of Thermodynamics is $\Delta U = Q + W$. The last chapter focused on W, the work done on a system, so now let's look more carefully at Q, the heat added to a system.

4.1   Heat Capacity

To find the heat added to a system we often use heat capacity, C, which is roughly defined as the amount of heat energy required to raise the temperature of a system by one degree. More precisely

$$C_V = \left(\frac{dQ}{dT}\right)_{V,\mathbf{M}} \qquad \text{and} \qquad C_p = \left(\frac{dQ}{dT}\right)_{p,\mathbf{M}}$$

are the heat capacities for a system at constant volume and at constant pressure. Heat capacity is an extensive variable; the specific heat capacities, $c_V = C_V/M$ and $c_p = C_p/M$, are intensive variables. From the definition of heat capacity

$$dQ = C_V dT \quad \text{(fixed } V, \mathbf{M}) \qquad \text{or} \qquad dQ = C_p dT \quad \text{(fixed } p, \mathbf{M}).$$

In general heat capacity is not a constant, it can depend on temperature, pressure, and other thermodynamic variables. That said, in many cases it is approximated as being constant (e.g., for water $c_V \approx c_p \approx 1$ calorie per gram per degree).

Energy and Heat Capacity

Using the First Law we can relate internal energy and heat capacity. Recall from calculus that for a function of three variables, $f(x, y, z)$, the differential of the function may be written as

$$df = \left(\frac{\partial f}{\partial x}\right)_{y,z} dx + \left(\frac{\partial f}{\partial y}\right)_{x,z} dy + \left(\frac{\partial f}{\partial z}\right)_{x,y} dz.$$

For a pure system, given $U(T, V, M)$ we can write the differential

$$dU = \left(\frac{\partial U}{\partial T}\right)_{V,M} dT + \left(\frac{\partial U}{\partial V}\right)_{T,M} dV + \left(\frac{\partial U}{\partial M}\right)_{T,V} dM.$$

Let's take a closed system (so $dM = 0$) which is at fixed volume ($dV = 0$) so

$$dU = \left(\frac{\partial U}{\partial T}\right)_{V,M} dT. \qquad \text{(fixed } V, M)$$

But we know that when volume is fixed $dW = -pdV = 0$ so by the First Law $dU = dQ$. This means that

$$C_V = \left(\frac{dQ}{dT}\right)_{V,M} = \left(\frac{\partial U}{\partial T}\right)_{V,M}.$$

In brief, at constant volume no work is done on the system so the change of internal energy equals the heat added.

Now let's consider the case where pressure is fixed instead of volume. Using the First Law

$$dU = dQ + dW = dQ - pdV$$

Again we consider a pure closed system so for fixed M

$$dQ = dU + pdV = \left\{ \left(\frac{\partial U}{\partial T}\right)_{V,M} dT + \left(\frac{\partial U}{\partial V}\right)_{T,M} dV \right\} + pdV$$

Using the result we just found for C_V

$$dQ = C_V dT + \left[\left(\frac{\partial U}{\partial V}\right)_{T,M} + p \right] dV \qquad \text{(fixed } M).$$

To understand the terms on the right hand side imagine the process that occurs when heat energy is added to a system held at constant pressure. Some of that energy goes towards raising the temperature; that's the first term. But since the volume changes, if U depends on V then some of the energy goes there (e.g., intermolecular potential energy); that's the second term. Finally, as the volume changes the system is doing work (e.g., raising a piston) and some of the energy goes into doing that work; that's the third term.

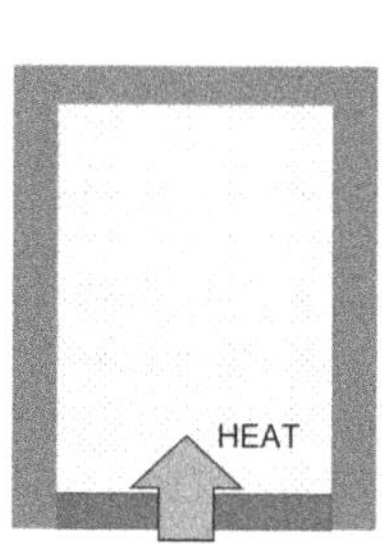
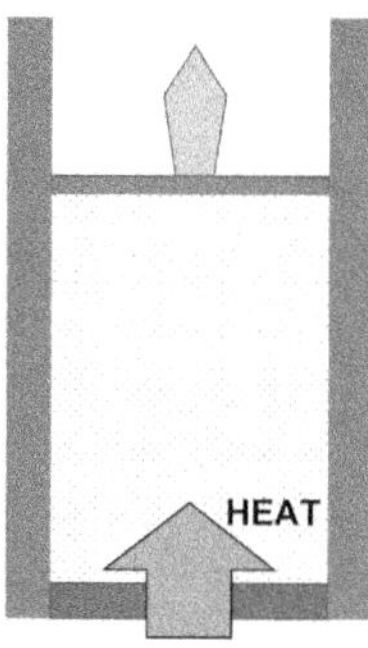

Figure 4.1: Adding heat to raise the temperature of a system held at constant volume (left) and constant pressure (right). Extra heat is needed when the system does work lifting the piston.

Using the expression above for dQ the heat capacity at constant pressure is

$$C_p = \left(\frac{dQ}{dT}\right)_{p,M} = C_V + \left[\left(\frac{\partial U}{\partial V}\right)_{T,M} + p\right]\left(\frac{\partial V}{\partial T}\right)_{p,M}$$

$$= C_V + \alpha V\left[\left(\frac{\partial U}{\partial V}\right)_{T,M} + p\right]$$

where α is the coefficient of thermal expansion (see Section 2.2).

Example 4.1: Given that for a pure ideal gas $U(T, V, M)$ is independent of volume show that $c_p - c_V = k_B/m$.
Solution: Since $U(T, V, M)$ is independent of volume $(\partial U/\partial V)_{T,M} = 0$. Furthermore, for an ideal gas, $\alpha = 1/T$, so

$$C_p = C_V + \frac{pV}{T} = C_V + N\mathrm{R} \qquad \text{(pure ideal gas)}$$

and $\hat{c}_p = \hat{c}_v + \mathrm{R}$. Since $\mathrm{R} = N_A k_B$ and $M = m\mathcal{N} = m N_A N$ we have

$$c_p = c_V + \frac{k_B}{m} \qquad \text{(pure ideal gas)}.$$

Notice that it takes more heat energy to raise the temperature when the pressure is constant than when the volume is constant; for an ideal gas that additional energy goes into the work done by the gas since it expands as the temperature increases (see Figure 4.1).

Example 4.2: Find $\hat{c}_p - \hat{c}_V$ for a pure van der Waals gas in terms of the coefficients a and b. Check that your answer gives $\hat{c}_p - \hat{c}_V = \mathrm{R}$ when $a = 0$.

Solution: For a van der Waals gas

$$\left(\frac{\partial U}{\partial V}\right)_{T,M} = \left(\frac{\partial}{\partial V}\right)_{T,M}\left[\hat{c}_V NT - a\frac{N^2}{V}\right] = a\frac{N^2}{V^2}.$$

Since $(\partial V/\partial T)_{p,M} = 1/\left(\partial T/\partial V\right)_{p,M}$, after some algebra

$$\left(\frac{\partial V}{\partial T}\right)_{p,M} = N\mathrm{R}\left(p - a\frac{N^2}{V^2}\frac{V - 2Nb}{V}\right)^{-1}.$$

Collecting terms gives

$$\hat{c}_p - \hat{c}_V = \frac{1}{N}\left[p + \left(\frac{\partial U}{\partial V}\right)_T\right]\left(\frac{\partial V}{\partial T}\right)_p = \mathrm{R}\left[p + a\frac{N^2}{V^2}\right]\left(p - a\frac{N^2}{V^2}\frac{V - 2Nb}{V}\right)^{-1}.$$

By inspection we get the ideal gas result of $\hat{c}_p - \hat{c}_V = \mathrm{R}$ when $a = 0$.

In Chapter 13 we'll derive this general relation:

$$C_p = C_V + \frac{TV\alpha^2}{\kappa_T}$$

For liquids and solids $C_p \approx C_V$ for typical values of α and κ_T.

4.2 Enthalpy

The enthalpy of a system, H, is defined as

$$H = U + pV$$

The First Law states that $\Delta U_{\mathrm{AB}} = W_{\mathrm{AB}} + Q_{\mathrm{AB}}$. When pressure is constant $W_{\mathrm{AB}} = -p\Delta V_{\mathrm{AB}}$ so

$$Q_{\mathrm{AB}} = \Delta U_{\mathrm{AB}} + p\Delta V_{\mathrm{AB}} = \Delta H_{\mathrm{AB}} \qquad \text{(fixed } p\text{)}.$$

That is, for a quasi-static process occurring at constant pressure the heat added equals the change in enthalpy. The symbol H is used for enthalpy because the name comes from the Greek word meaning "to heat."

Schroeder* has a whimsical example to explain the difference between H and U:

> To create a rabbit out of nothing and place it on the table, the magician must summon up not only the energy U of the rabbit, but also some additional energy, equal to pV, to push the atmosphere out of the way to make room. The *total* energy required is the **enthalpy.**

---

*Daniel V. Schroeder, *An Introduction to Thermal Physics*, Oxford Univ. Press (2021).

Example 4.3: One mole of magnesium carbonate ($MgCO_3$) solid absorbs 26 kilocalories of heat when it is decomposed into magnesium oxide (MgO) solid and carbon dioxide (CO_2) gas at 900 Kelvin and atmospheric pressure ($p = 1.01 \times 10^5$ Pa). Find the change in internal energy for this process at constant pressure.

Solution: Looking up the mass densities of magnesium carbonate and magnesium oxide in the solid phase we find they have molar volumes of $V/N = 0.028$ and 0.011 liters per mole, respectively. From the ideal gas law the carbon dioxide in the gas phase has a molar volume of $V/N = RT/p = 74$ liters per mole. For this process $Q_{AB} = \Delta H_{AB} = 26$ kcal $= 1.09 \times 10^5$ Joules. The change in internal energy is

$$\Delta U_{AB} = \Delta H_{AB} - p\Delta V_{AB} = (1.09 \times 10^5 \text{ J}) - (1.01 \times 10^5 \text{ Pa})(0.074 \text{ m}^3) = 1.01 \times 10^5 \text{ J}.$$

Notice that the volume of the solid magnesium compounds is negligible compared with the volume of the mole of carbon dioxide gas that's produced. We see that about 7% of the heat absorbed in this process goes into pushing the atmosphere out of the way as the CO_2 is created.

As you know, Q_{AB} depends on the path going from state A to state B; when that process is isobaric (constant pressure) then $Q_{AB} = \Delta H_{AB}$. For a process that's *not* isobaric we can determine $H_{AB} = H_B - H_A$ from the initial and final states but the change in enthalpy won't be equal to the heat added for that process. My point is that enthalpy is always defined but mostly useful when pressure is constant.

Enthalpy and Heat Capacity

Since the heat added at constant pressure equals the change of enthalpy (i.e., $dQ = dH$) we have

$$C_p = \left(\frac{dQ}{dT} \right)_{p,M} = \left(\frac{\partial H}{\partial T} \right)_{p,M}$$

so $dH = C_p dT$ when pressure and mass are fixed. Compare this with

$$C_V = \left(\frac{dQ}{dT} \right)_{V,M} = \left(\frac{\partial U}{\partial T} \right)_{V,M}$$

so $dU = C_V dT$ when volume and mass are fixed. As we've seen $C_p \geq C_V$ so think of it as dU being the bill and dH being what you pay including the tip (i.e., the work, pdV, done by the system).

4.3 Adiabatic Processes

Many quasi-static processes are defined by constraints, such as isothermal processes that hold temperature fixed and isobaric processes that hold the pressure fixed. Another important class of quasi-static processes are *adiabatic processes* for which $Q_{AB} = 0$. By the First Law this means that in going from state A to state B the change of internal energy is entirely due to work, that is, $\Delta U_{AB} = W_{AB}$. As we'll see later, adiabatic processes are related to the constraint of holding entropy fixed.

Adiabatic Processes in an Ideal Gas

Consider a pure closed, adiabatic system, that is, a system with fixed M (closed) and with $dQ = 0$ (adiabatic). From earlier results

$$dQ = C_V dT + \left[\left(\frac{\partial U}{\partial V}\right)_{T,M} + p\right] dV = 0.$$

Furthermore, for a pure ideal gas U does not depend on V so this simplifies to

$$c_V M dT + p dV = 0 \qquad \text{(adiabatic, ideal gas)}.$$

Using the ideal gas law

$$c_V dT + \frac{k_B T}{mV} dV = 0.$$

In an example above we showed that for an ideal gas $c_p - c_V = k_B/m$ so

$$\frac{dT}{T} = -\frac{c_p - c_V}{c_V}\frac{dV}{V} = -(\gamma - 1)\frac{dV}{V}$$

where $\gamma = c_p/c_V$ is the ratio of the heat capacities. Finally, integrating both sides gives

$$TV^{(\gamma - 1)} = \text{constant} \qquad \text{(adiabatic, ideal gas)}$$

where the constant is a constant of integration.

Example 4.4: Figure 4.2 illustrates the adiabatic expansion of an ideal gas. For a pure ideal diatomic gas ($c_V = \frac{5}{2} k_B/m$) at an initial temperature of $T_A = 300$ K find the final temperature if the gas expands adiabatically to twice its initial volume (i.e., $V_B = 2V_A$).
Solution: Given $c_V = \frac{5}{2} k_B/m$

$$c_p = c_V + \frac{k_B}{m} = \frac{7}{2}\frac{k_B}{m}$$

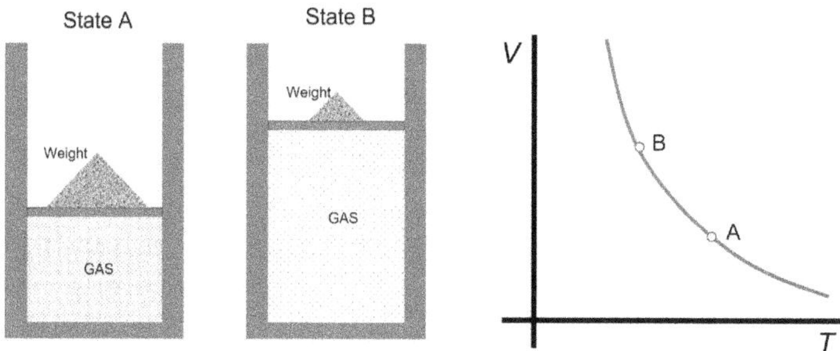

Figure 4.2: Temperature decreases as volume increases in the adiabatic expansion of an ideal gas.

so $\gamma = c_p/c_V = 7/5$. From the above $TV^{(\gamma-1)}$ must be the same at both states, that is,

$$T_B V_B^{(\gamma-1)} = T_A V_A^{(\gamma-1)}$$

or

$$T_B = T_A(V_A/V_B)^{(\gamma-1)} = T_A(1/2)^{2/5}.$$

The final temperature is thus $T_B = 227$ K. No heat enters or leaves the system yet the gas temperature goes down because the gas loses internal energy in doing work as it expands. Note that this expansion is *not* at constant pressure.

Using the ideal gas law you can show that for an adiabatic process from state A to state B

$$p_B V_B^\gamma = p_A V_A^\gamma \qquad \text{and} \qquad T_B^\gamma p_B^{-(\gamma-1)} = T_A^\gamma p_A^{-(\gamma-1)}.$$

Example 4.5: Find the work done on an ideal gas in an adiabatic compression expressed in terms of the initial and final temperatures.
Solution: Using the results above

$$W_{AB} = -\int_{V_A}^{V_B} p(V)dV = -C\int_{V_A}^{V_B} V^{-\gamma}dV = \frac{C}{\gamma-1}(V_B^{-\gamma+1} - V_A^{-\gamma+1}).$$

The constant may be written as $C = p_A V_A^\gamma = p_B V_B^\gamma$ so

$$W_{AB} = \frac{1}{\gamma-1}(p_B V_B - p_A V_A) = \frac{NR}{\gamma-1}(T_B - T_A) = \hat{c}_V N(T_B - T_A).$$

This result confirms that $W_{AB} = \Delta U_{AB}$ since $Q_{AB} = 0$.

Adiabatic processes typically occur when the compression or expansion is rapid enough that heat conduction is negligible while being slow enough to be quasi-static. An important example is the rapid expansion and contraction that occurs in sound waves (e.g., audible frequencies are typically in the range of 100 Hz to 10,000 Hz).

$$* * *$$

At this point the formulation of heat is incomplete by comparison with that of work. For the latter we have the general result that $dW = -pdV$ while for heat we have formulated dQ only for special cases, such as isochoric, isobaric, or adiabatic processes. This imbalance is removed in the next chapter by the introduction of the concept of entropy.

Chapter 5

Entropy

Many textbooks introduce entropy and the Second Law of Thermodynamics by following the historical steps by which they were discovered. Other authors use a formal, axiomatic approach. We'll use "reverse engineering", first defining entropy without justification and in later chapters show how it fits with the theories developed by Carnot, Clausius, and Gibbs.

5.1 Entropy and Heat

Recall that the First Law of Thermodynamics can be written as

$$dU = dQ + dW \qquad \text{or} \qquad \Delta U_{\text{AB}} = Q_{\text{AB}} + W_{\text{AB}}.$$

We also have an expression for work in terms of state variables

$$dW = -pdV \qquad \text{or} \qquad W_{\text{AB}} = -\int_{V_{\text{A}}}^{V_{\text{B}}} p(V)dV$$

where W_{AB} is the work done *on* the system in going from state A to B along some given path (i.e., for a given $p(V)$).

It would be convenient to have a result for dQ similar to what we have for dW. We do have $dQ = CdT$ but the heat capacity depends on the process ($C = C_V$ for isochoric processes, $C = 0$ for adiabatic processes, etc.). What we need is a new state variable and it turns out that entropy, S, is the missing piece. For closed systems

$$dQ = Td_eS \qquad \text{or} \qquad Q_{\text{AB}} = \int_{S_{\text{A}}}^{S_{\text{B}}} T(S)d_eS \qquad \text{(Closed system)}$$

where Q_{AB} is the heat added to the system. For open systems we also have to include a term that accounts for mass entering or leaving the system. That

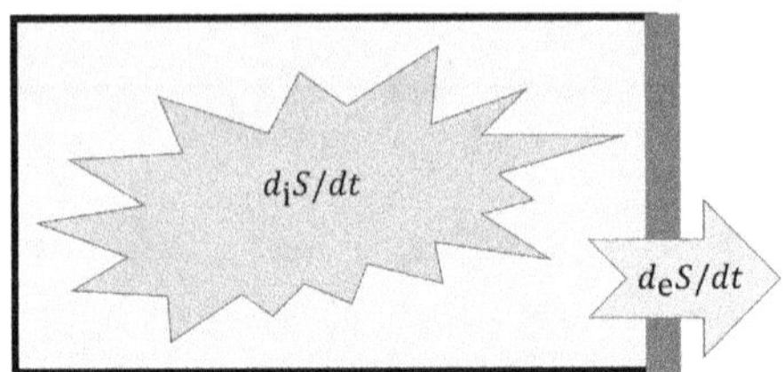

Figure 5.1: System with both internal (e.g., chemical reactions) and external (e.g., heat flow) forms of entropy change.

contribution is derived in Chapter 9 but for now we'll just consider closed systems. By the way, notice that entropy is an extensive state variable that has units of Joules per Kelvin, which are also the units of the Boltzmann constant, k_B, and of heat capacity.

In modern thermodynamics the subscript "e" on the $d_e S$ is used to remind us that this is a change of entropy that occurs from an *external* source, specifically, due to a transfer of heat energy into the system. If this is the only way that entropy changes then $dS = d_e S$. In general the total entropy change is $dS = d_e S + d_i S$ where $d_i S$ is the change due to internal sources, such as material transformations. For example, Figure 5.1 illustrates a system in which a chemical reaction occurs ($d_i S/dt > 0$) while at the same time energy, in the form of heat, is leaving the system ($d_e S/dt < 0$).

Calculating Entropy

We can find the change in entropy of a system by using calorimetry to measure experimentally the heat added, Q_{AB}, in going between states. In closed, homogeneous systems without internal entropy production (e.g., no chemical reactions) $d_i S = 0$ so $dS = d_e S$. The change in entropy is given by

$$\Delta S_{AB} = S_B - S_A = \int_A^B dS = \int_A^B \frac{dQ}{T} \qquad \text{(closed, inert)}.$$

As with internal energy we can define a reference state for entropy, S_0, and write,

$$S = S_0 + \int_0 dS = S_0 + \int_0 \frac{dQ}{T}.$$

Usually we're only interested in the change in entropy, ΔS_{AB}, which is independent of the choice of reference state.

The differentials $d_e S$ and dS are exact (like dU and dV) so if we select a quasi-static path between A and B such that $T = T^{\text{res}}$ is fixed by a reservoir

$$\Delta S_{AB} = \int_A^B \frac{dQ}{T^{\text{res}}} = \frac{1}{T^{\text{res}}} \int_A^B dQ = \frac{Q_{AB}}{T^{\text{res}}} \qquad \text{(Isothermal process)}.$$

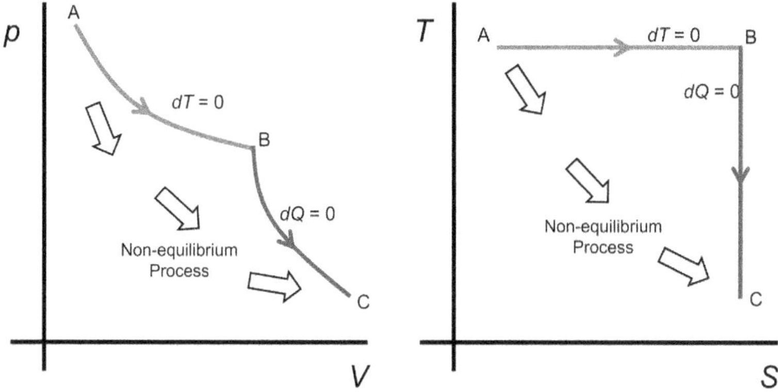

Figure 5.2: Pressure-Volume (left) and Temperature-Entropy (right) diagrams for two processes going from state A to state C for a closed, inert system (**M** fixed). Path ABC combines an isothermal branch (AB) and an adiabatic branch (BC). The non-equilibrium process has no path since pressure and temperature are undefined during the process. The change in entropy, $\Delta S_{AC} = S_C - S_A$, is the same for these two processes because entropy is a state variable.

But what if we cannot get from A to B along an isothermal path? Along an adiabatic path (no heat added or removed) $dQ = 0$ so $d_e S = 0$ and $\Delta S = 0$. In general, we can build a path between any two states A and B by combining a sequence of isothermal and adiabatic paths and thus calculate ΔS (see Figure 5.2).

Example 5.1: One kilogram of liquid water at $20°$ C is brought into contact with a large heat reservoir at $80°$. What is the change in entropy of the water and of the reservoir once they reach equilibrium? Assume this to be a quasi-static process at constant pressure.

Solution: Call state A the initial state ($T_A = 293$ K), state C the final state ($T_C = 353$ K). The change of entropy of the water going from state A to C is

$$\Delta S_{\text{water}} = S_C - S_A = \int_A^C d_e S = \int_A^C \frac{dQ}{T} = \int_A^C \frac{C_p dT}{T} = C_p \ln(T_C/T_A) = 779 \text{ J/K}$$

with $C_p = 4180$ J/K for one kilogram of water.

The total heat added to the water is $Q_{AC} = C_p(T_C - T_A) = 2.51 \times 10^5$ J and by First Law this is also the amount of heat energy removed from the reservoir. Since the reservoir is at fixed temperature $T^{\text{res}} = T_C$ then $\Delta S_{\text{res}} = -Q_{AC}/T_C = -710$ J/K. For the water and reservoir combined we have a net increase of entropy of $\Delta S_\Sigma = \Delta S_{\text{water}} + \Delta S_{\text{res}} = 69$ J/K.

Example 5.2: Repeat the previous example but heat the water from 20° C to 80° by first bringing it in contact with a reservoir at 50° C and then with the reservoir at 80° C.

Solution: Call state B the point when the water reaches the temperature of the first reservoir ($T_B = 323$ K). The change of entropy of the water is the same since

$$\Delta S_{\text{water}} = \int_A^B \frac{C_p dT}{T} + \int_B^C \frac{C_p dT}{T} = C_p \ln(T_C/T_A) = 779 \text{ J/K}.$$

For the first reservoir we have $\Delta S_{\text{res1}} = -Q_{AB}/T_B = -388$ J/K. For the second reservoir we have $\Delta S_{\text{res2}} = -Q_{BC}/T_C = -355$ J/K. By using the intermediate reservoir the net increase of entropy is reduced from $\Delta S_\Sigma = 69$ J/K to 36 J/K.

Example 5.3: Find ΔS_{AB} for an isothermal expansion of a pure ideal gas from volume V_A to V_B with $T_A = T_B = T^{\text{res}}$ (see Figure 3.3).

Solution: From the examples in Section 3.2

$$Q_{AB} = -W_{AB} = \int_A^B p(V)\, dV = \int_A^B \frac{NRT^{\text{res}}}{V}\, dV = NRT^{\text{res}} \ln(V_B/V_A) > 0.$$

Heat enters the system from the reservoir to maintain it at constant temperature as the gas expands. The change of entropy for the gas is

$$\Delta S_{AB} = \frac{Q_{AB}}{T^{\text{res}}} = NR\ln(V_B/V_A) = NR\ln(V_B) - NR\ln(V_A).$$

Writing entropy as a function of T, V, and N it must be that for an ideal gas

$$S(T,V,N) = NR\ln(V) + f(T,N) \qquad \text{(Ideal gas)}$$

where $f(T,N)$ is a function that, as we'll later see, depends on heat capacity.

Theoretical modeling is an alternative to experimental measurements for calculating entropy. For example, using the theory of ensembles from statistical mechanics one may formulate $S(U,V,M)$ or $S(T,V,M)$ from a microscopic model of the system. Experiments can validate these theoretical models and relate parameters to physical quantities, as in the next example.

Example 5.4: The theory of ensembles derives the entropy for a certain system to be

$$S(T,V,M) = MC\ln T + f(V,M)$$

where C is a constant parameter. Find the heat added by a going from T_A to T_B while holding V and M fixed. For this system $d_i S = 0$ so entropy change is $dS = d_e S = dQ/T$.

Solution: First, we'll write

$$T(S, V, M) = \exp\left(\frac{S - f}{M\mathcal{C}}\right).$$

Since V and M are fixed f is a constant. The heat added is

$$Q_{\text{AB}} = \int_{S_{\text{A}}}^{S_{\text{B}}} T(S)\,dS = \left[M\mathcal{C}\exp\left(\frac{S - f}{M\mathcal{C}}\right)\right]_{S_{\text{A}}}^{S_{\text{B}}} = M\mathcal{C}(T_{\text{B}} - T_{\text{A}}).$$

From this result we see that $\mathcal{C} = c_V$ is the specific heat capacity at fixed volume.

Finally, the entropy may be formulated from knowing other thermodynamic quantities, such as the equations of state. This approach is illustrated in the next section to derive the entropy of an ideal gas.

5.2   Entropy of an Ideal Gas

In an earlier example we considered an isothermal expansion of a pure ideal gas and found that the entropy can be written as

$$S(T, V, M) = M\frac{k_{\text{B}}}{m}\ln(V) + f(T, M)$$

since $N R = k_{\text{B}} M/m$. Now let's find $f(T, M)$.

From multivariate calculus the differential of a function of two variables is

$$df(x, y) = \left(\frac{\partial f}{\partial x}\right)_y dx + \left(\frac{\partial f}{\partial y}\right)_x dy.$$

Holding M fixed

$$dS(T, V) = \left(\frac{\partial S}{\partial T}\right)_{V,M} dT + \left(\frac{\partial S}{\partial V}\right)_{T,M} dV.$$

From our earlier results

$$\left(\frac{\partial S}{\partial V}\right)_{T,M} = \left(\frac{\partial}{\partial V}\right)_{T,M}\left\{M\frac{k_{\text{B}}}{m}\ln(V) + f(T, M)\right\} = M\frac{k_{\text{B}}}{m}\frac{1}{V}$$

and

$$\left(\frac{\partial S}{\partial T}\right)_{V,M} = \frac{1}{T}\left(\frac{dQ}{dT}\right)_{V,M} = \frac{c_V M}{T}.$$

Collecting the above

$$dS = \frac{c_V M}{T}dT + M\frac{k_{\text{B}}}{m}\frac{1}{V}dV.$$

Assuming c_V is constant, integrating from a reference state, T_0 and V_0, to T and V we find

$$S(T,V,M) - S(T_0,V_0,M) = c_V M \ln(T/T_0) + M\frac{k_B}{m}\ln(V/V_0)$$

or

$$S(T,V,M) = Mc_V \ln T + M\frac{k_B}{m}\ln V + S_0(M)$$

where S_0 depends on the reference state.

We can go even further by noting that entropy is an extensive function. That is, if we double V and M while holding T fixed then S must double. The more elegant way to say this is that for an arbitrary positive constant, $\mathcal{C}$, since entropy is extensive

$$S(T,\mathcal{C}V,\mathcal{C}M) = \mathcal{C}S(T,V,M).$$

By contrast, temperature is an intensive function so

$$T(\mathcal{C}U,\mathcal{C}V,\mathcal{C}M) = T(U,V,M).$$

See Table 1.1 for more examples of extensive and intensive functions. From the fact that entropy is extensive you can show that,

$$S(T,V,M) = M\left[s_0 + \frac{k_B}{m}\ln(V/M) + c_V\ln(T)\right] \qquad \text{(Ideal Gas)}$$

where s_0 is a constant.

Example 5.5: Find the change of entropy when N moles of a pure ideal gas goes from temperature T_A to T_B at constant pressure p_0.
Solution: From the ideal gas law $V = NRT/p_0$ so

$$\begin{aligned}
S(T,V,M) &= M\left[s_0 + \frac{k_B}{m}\ln\left(\frac{NRT}{Mp_0}\right) + c_V\ln(T)\right] \\
&= M\left[\left\{s_0 + \frac{k_B}{m}\ln\left(\frac{NR}{Mp_0}\right)\right\} + \frac{k_B}{m}\ln(T) + c_V\ln(T)\right]
\end{aligned}$$

where $M = mNN_A$. For a pure ideal gas $c_p = c_V + k_B/m$ (see Example in Section 4.1) so

$$S(T,p_0,M) = Mc_p\ln(T) + f(p_0,M).$$

The change of entropy in going from T_A to T_B at constant pressure is

$$\Delta S_{AB} = S(T_B,p_0,M) - S(T_A,p_0,M) = N\hat{c}_p\ln(T_B/T_A)$$

since $Mc_p = N\hat{c}_p = C_p$. The same result is obtained from $d_eS = dQ/T$ and the fact that $dQ = C_p dT$ when pressure is constant. Note that the entropy of the gas increases if the temperature increases.

5.3 Entropy and Phase Transformations*

The entropy change during a phase transformation, such as solid ice melting into liquid water, is relatively simple since the process occurs at constant temperature. We usually consider this as occurring at constant pressure (e.g., at atmospheric pressure) so $Q = \Delta U + p\Delta V = \Delta H$. The entropy of fusion (freezing/melting) is

$$\Delta S_{\text{fus}} = \frac{Q_{\text{fus}}}{T_{\text{melt}}} = \frac{\Delta H_{\text{fus}}}{T_{\text{melt}}}$$

where T_{melt} is the melting temperature and ΔH_{fus} is the enthalpy of fusion. Similarly, the entropy of vaporization (condensing/boiling) is

$$\Delta S_{\text{vap}} = \frac{Q_{\text{vap}}}{T_{\text{boil}}} = \frac{\Delta H_{\text{vap}}}{T_{\text{boil}}}$$

where T_{boil} is the boiling temperature and ΔH_{vap} is the enthalpy of vaporization. The heat exchanged in these processes is easily measured by calorimetry and all these quantities are tabulated on-line for many materials. For example, the specific enthalpy of fusion (also known as latent heat of fusion) for water is about 80 calories per gram.

Example 5.6: Compare the change of entropy of one kilogram of liquid nitrogen boiling into gas with the change of entropy for that gas heating to a room temperature of 300 K. In both cases the processes occur at atmospheric pressure; treat the nitrogen gas as a classical diatomic gas ($f_{\text{d}} = 5$).
Solution: For nitrogen $\Delta H_{\text{vap}}/M = 199$ kJ/kg and $T_{\text{boil}} = 77.4$ K so $\Delta S_{\text{vap}} = 2.57$ kJ/K. The change of entropy in heating the gas from $T_{\text{A}} = T_{\text{boil}}$ to room temperature is

$$\Delta S_{\text{AB}} = \int_{T_{\text{boil}}}^{T_{\text{B}}} d_e S = \int_{T_{\text{A}}}^{T_{\text{B}}} \frac{dQ}{T} = \int_{T_{\text{A}}}^{T_{\text{B}}} \frac{C_p dT}{T} = N\hat{c}_p \ln(T_{\text{B}}/T_{\text{A}}) \qquad \text{(constant } p\text{)}.$$

Notice that this is the same result as obtained in the previous example. For $M = 1$ kg of nitrogen gas (N_2), $N = M/(28 \text{ g/mol}) = 35.7$ mol. The molar heat capacity of a clasical diatomic gas is $\hat{c}_p = \hat{c}_V + R = \frac{7}{2}R$ so $\Delta S_{\text{AB}} = 1.41$ kJ/K. The reason that ΔS_{vap} is nearly twice as large as ΔS_{AB} is that there's a large increase in entropy associated with the increase in volume when a liquid transforms into a gas.

$$* \ * \ *$$

Now that we've defined entropy we're almost ready to formulate the Second Law of Thermodynamics. Before doing so let's examine the motivation behind this law by analyzing the difference between reversible and irreversible processes.

Chapter 6

Reversible and Irreversible Processes

In this chapter we'll use entropy to define the difference between reversible and irreversible processes as a prelude to formulating the Second Law of Thermodynamics.

6.1 Irreversible Processes

Chapter 5 introduced entropy by relating it to heat added to a closed system as $dQ = Td_eS$. For homogeneous, inert systems $dS = d_eS$ and

$$\Delta S_{\mathrm{AB}} = \int_{\mathrm{A}}^{\mathrm{B}} \frac{1}{T}\, dQ.$$

This path integral is easy for two special cases: First, for an adiabatic process $dQ = 0$ so $\Delta S_{\mathrm{AB}} = Q_{\mathrm{AB}} = 0$. Second, for an isothermal process

$$\Delta S_{\mathrm{AB}} = \frac{1}{T^{\mathrm{res}}} \int_{\mathrm{A}}^{\mathrm{B}} dQ = \frac{Q_{\mathrm{AB}}}{T^{\mathrm{res}}}$$

where T^{res} is a fixed reservoir temperature. Let's use these relations to see how entropy changes in an irreversible process.

Energy and Entropy in Irreversible Processes

Let's start with a simple example of a quasi-static irreversible process. Our system is isolated, inert and heterogeneous with three parts: a solid rock plus a pair of thermal reservoirs at temperatures T^{cold} and T^{hot} with $T^{\mathrm{cold}} < T^{\mathrm{hot}}$.

Consider the cyclic process in which the rock is moved between the two reservoirs (see Figure 6.1). The rock is initially cold with temperature $T_{\mathrm{A}} =$

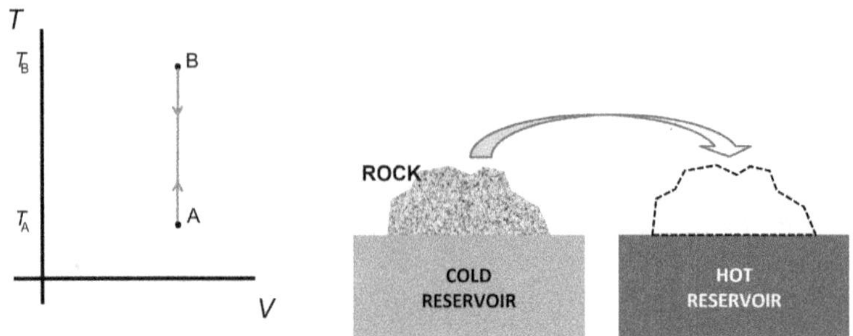

Figure 6.1: Irreversible, quasi-static, isochoric (fixed V) process between temperatures T_A and T_B for a rock that is alternatingly brought into contact with hot and cold reservoirs.

T^{cold} (state A). It is put into thermal contact with the hot reservoir and its temperature slowly (quasi-statically) increases until it reaches temperature $T_B = T^{\mathrm{hot}}$ (state B). Then the rock is placed in contact with the cold reservoir and allowed to quasi-statically return to its initial state. The process occurs at constant pressure and we assume that the volume of the rock stays constant (i.e., the coefficient of thermal expansion $\alpha \approx 0$).

Example 6.1:   For the process illustrated in Figure 6.1 find Q_{AB} and Q_{BA} for the rock and for each reservoir; also find ΔU for one complete cycle. The mass of the rock is 100 g and its specific heat capacity is $c_V = 800$ J/(kg K); the reservoir temperatures are $T^{\mathrm{hot}} = 400$ K and $T^{\mathrm{cold}} = 300$ K.
Solution: The heat added to the rock in going from state A to state B is

$$Q_{AB}^{\mathrm{rock}} = \Delta U_{AB}^{\mathrm{rock}} = Mc_V(T_B - T_A) = 8000 \text{ J}.$$

This heat comes from the hot reservoir so $Q_{AB}^{\mathrm{hot}} = -Q_{AB} = -8000$ J while $Q_{AB}^{\mathrm{cold}} = 0$. In taking the rock from state B back to state A we have $Q_{BA} = -8000$ J, $Q_{BA'}^{\mathrm{hot}} = 0$, and $Q_{BA'}^{\mathrm{cold}} = 8000$ J. Notice that we call A$'$ the final state for the reservoirs since their total energy changes.
    The total change of internal energy (rock plus reservoirs) for the full cycle is

$$\Delta U_{\circlearrowleft}^{\Sigma} = \Delta U_{ABA}^{\mathrm{rock}} + \Delta U_{ABA'}^{\mathrm{hot}} + \Delta U_{ABA'}^{\mathrm{cold}} = (0) + (-8000 \text{ J}) + (+8000 \text{ J}) = 0,$$

as expected from energy conservation. Note that this result does *not* require the process to occur quasi-statically since internal energy is a state variable (i.e., ΔU is path independent).

Now let's analyze this example using entropy. The significant result will be that for the hot reservoir $\Delta S^{\mathrm{hot}} < 0$ and for the cold reservoir $\Delta S^{\mathrm{cold}} > 0$. But the entropy lost by the hot reservoir is *not* equal to the entropy gained by the cold reservoir!

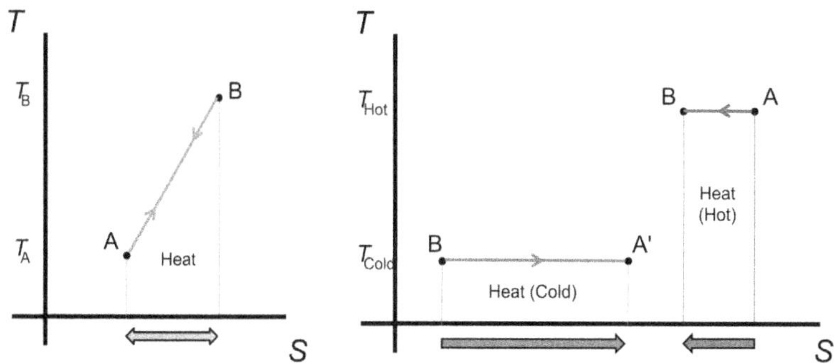

Figure 6.2: The $T-S$ diagrams for a rock (left) and a pair of reservoirs (right); see Fig. 6.1. The rock's entropy increases (path A $\rightarrow$ B) and decreases (path B $\rightarrow$ A) by contact with the hot and cold reservoirs. The reservoir temperatures, T^{hot} and T^{cold}, are constant but their entropies change. *Important:* The areas under paths are equal since $|Q^{\text{rock}}_{\text{AB}}| = |Q^{\text{rock}}_{\text{BA}}| = |Q^{\text{hot}}_{\text{AB}}| = |Q^{\text{cold}}_{\text{BA}'}|$.

Example 6.2:   Continuing the previous example, find the change in entropy of the rock and the reservoirs after one full cycle.

Solution: The rock returns to its initial state so $\Delta S^{\text{rock}}_{\text{ABA}} = 0$ since entropy is a state variable. The reservoir temperatures are constant and for an isothermal process

$$\Delta S^{\text{hot}}_{\text{AB}} = \frac{Q^{\text{hot}}_{\text{AB}}}{T^{\text{hot}}} = \frac{-8000 \text{ J}}{400 \text{ K}} = -20 \text{ J/K},$$

$$\Delta S^{\text{cold}}_{\text{BA}'} = \frac{Q^{\text{cold}}_{\text{BA}'}}{T^{\text{cold}}} = \frac{+8000 \text{ J}}{300 \text{ K}} = +27 \text{ J/K}.$$

For the hot reservoir nothing happens going from state B to state A' so $\Delta S^{\text{hot}}_{\text{BA}'} = 0$ and for the cycle $\Delta S^{\text{hot}}_{\text{ABA}'} = -20 \text{ J/K}$. Similarly $\Delta S^{\text{cold}}_{\text{ABA}'} = +27 \text{ J/K}$ so the total change in entropy is

$$\Delta S^{\Sigma}_{\circlearrowleft} = \Delta S^{\text{rock}}_{\text{ABA}} + \Delta S^{\text{hot}}_{\text{ABA}'} + \Delta S^{\text{cold}}_{\text{ABA}'} = 7 \text{ J/K}.$$

Contrast this with $\Delta U^{\Sigma}_{\circlearrowleft} = 0$.

The fact that $\Delta S^{\Sigma} > 0$ is a general result for irreversible processes. Figure 6.2 illustrates the cycle on the $T - S$ diagrams for each subsystem (rock and the two reservoirs). The importance of this example is that although the rock returns to its original state this is *not* a reversible process because energy (as heat) is irreversibly transferred from the hot reservoir to the cold reservoir. The reverse cannot occur; we cannot use the rock to draw heat energy from the cold reservoir and transfer it to the hot reservoir. This intuitive result foreshadows the Second Law.

6.2   Reversible Processes

Adiabatic Processes

The example in the previous section analyzed an irreversible process in which the temperature of a rock was raised and then lowered. Now let's consider a different situation in which the temperature is also raised and lowered but by a reversible process, specifically, the adiabatic compression and expansion of an ideal gas. The gas is modeled as a closed, inert system with initial temperature T_A and volume V_A. By slowly increasing the pressure the gas is quasi-statically compressed to a volume V_B.

By definition, for an adiabatic process there is no heat added or removed (i.e., the gas container is thermally insulated) so $Q_{AB} = 0$. Nevertheless, the compression causes the temperature to increse, specifically, from the results derived in Section 4.3

$$T_B = T_A \left(\frac{V_A}{V_B} \right)^{\gamma - 1}$$

where $\gamma = c_p/c_V \geq 1$. Since $V_B < V_A$ then $T_B > T_A$. This temperature increase makes sense since by First Law $\Delta U_{AB} = Q_{AB} + W_{AB} > 0$, that is, the gas gains internal energy because work is done *on* the system by the compression (i.e., $W_{AB} > 0$).

This quasi-static process is reversible; we simply adiabatically decrease the pressure and expand the gas back to its original volume. Going from state B back to state A the work done on the system is $W_{BA} = -W_{AB}$ and, again, $Q_{BA} = 0$. For the cycle $W_{ABA} = 0$ and $Q_{ABA} = 0$. But, most importantly, since no heat is added or removed

$$\Delta S_{AB} = \Delta S_{BA} = \Delta S_{ABA} = 0.$$

And this result holds for all closed, inert homogeneous systems, not just for an ideal gas, undergoing an adiabatic process.

Isothermal Processes

Now consider an ideal gas undergoing an isothermal process in which the gas goes from volume V_A to V_B while held at constant temperature T^{res} by a reservoir. An example in Section 3.2 showed that the work done on the gas and the heat added to the gas are

$$W_{AB} = -Q_{AB} = NRT^{res} \ln(V_A/V_B).$$

Again, consider the case where this a compression ($V_B < V_A$) so $W_{AB} > 0$ and $Q_{AB} < 0$.

During this compression the gas maintains itself at constant temperature by losing energy (as heat) to the reservoir. The change of entropy for the gas

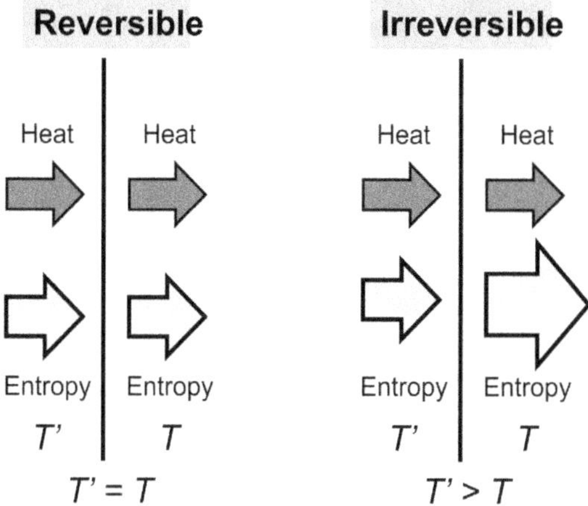

Figure 6.3: Energy (heat) and entropy transfer for reversible and irreversible processes. Energy is always conserved but entropy is conserved only for reversible processes.

and for the reservoir are

$$\Delta S_{\mathrm{AB}} = \frac{Q_{\mathrm{AB}}}{T^{\mathrm{res}}} < 0 \qquad \text{and} \qquad \Delta S_{\mathrm{AB}}^{\mathrm{res}} = \frac{Q_{\mathrm{AB}}^{\mathrm{res}}}{T^{\mathrm{res}}} > 0.$$

By First Law (i.e., conservation of energy) the heat energy removed from the gas must equal the heat energy added to the reservoir, that is, $Q_{\mathrm{AB}} = -Q_{\mathrm{AB}}^{\mathrm{res}}$ so

$$\Delta S_{\mathrm{AB}}^{\Sigma} = \Delta S_{\mathrm{AB}} + \Delta S_{\mathrm{AB}}^{\mathrm{res}} = 0.$$

For this isothermal compression the gas loses entropy but the reservoir gains an equal amount of entropy so the total change of entropy is zero.

Going from state B back to state A we have $\Delta S_{\mathrm{BA}} = -\Delta S_{\mathrm{AB}}$ and $S_{\mathrm{BA}}^{\mathrm{res}} = -\Delta S_{\mathrm{AB}}^{\mathrm{res}}$ so the entropy that was given to the reservoir is all returned back to the gas. The isothermal processes A $\to$ B and B $\to$ A are both reversible since the total (gas plus reservoir) change of entropy is zero. This result holds for all closed, inert systems undergoing an isothermal process with a reservoir, not just for an ideal gas. Figure 6.3 illustrates the difference between reversible and irreversible heat transfer.

Contrast this with the *free expansion* of a pure ideal gas in a heterogenous, isolated system (see Figure 6.4). Initially the gas is in the left chamber (volume V_{L}) with the right chamber (volume V_{R}) being vacuum. A hole is opened and the gas fills the entire system uniformly at equilibrium. Since the internal energy is independent of volume for an ideal gas the temperature remains constant. However this is an *irreversible* process; from Section 5.2

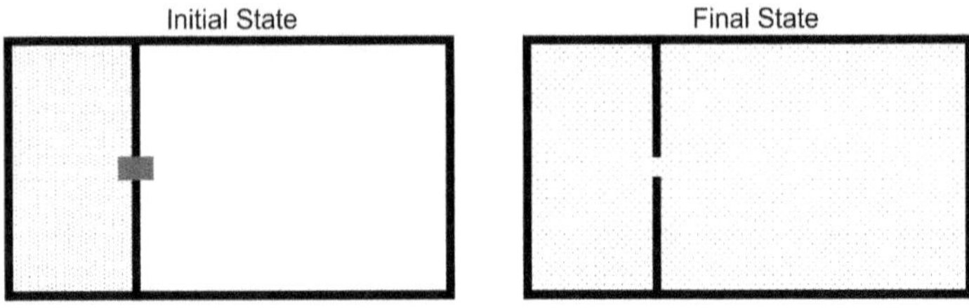

Figure 6.4: Free expansion of an ideal gas in a heterogenous, isolated system.

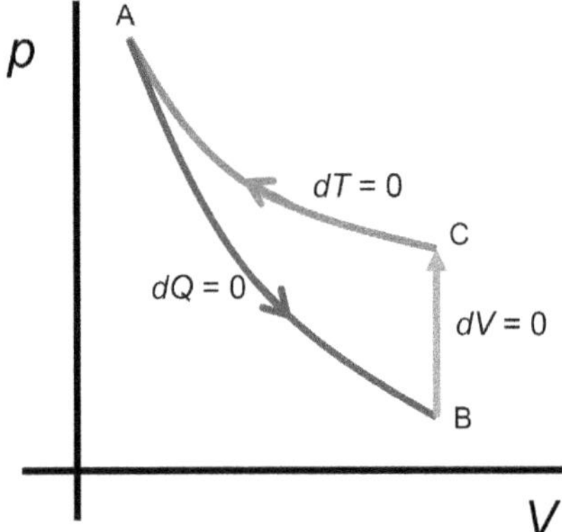

Figure 6.5: Cycle consisting of an adiabatic process A → B, an isochoric process B → C, and an isothermal process C → A.

the change in entropy is

$$\Delta S_{AB} = S(T, V_{L} + V_{R}, M) - S(T, V_{L}, M) = \frac{M k_{B}}{m} \ln \left(\frac{V_{L} + V_{R}}{V_{L}} \right) > 0.$$

Again, the key to distinguishing between reversible and irreversible processes is the total change of entropy.

Example 6.3: A dilute inert gas in a closed system undergoes the cycle illustrated in Figure 6.5. Starting from state A the gas pressure is lowered adiabatically, cooling the system. When it reaches state B the system is brought into contact with a reservoir and the volume is held fixed while the gas temperature increases to $T^{\mathrm{res}} = T_{A}$. Finally, the system is kept in contact with the reservoir, compressing the gas isothermally until it returns to state A. Show that this cycle is irreversible, that is, show that the total entropy of the gas plus the reservoir increases after one cycle.

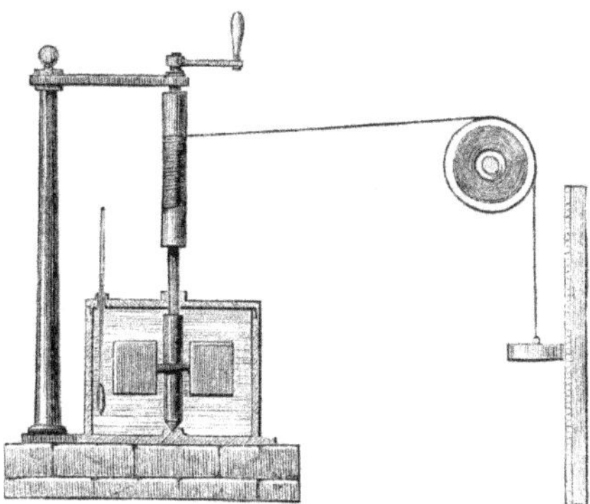

Figure 6.6: Joule's experiment demonstrating the mechanical equivalence of heat.

Solution: For this cycle

$$\Delta S_{\circlearrowleft}^{\Sigma} = \Delta S_{AB}^{gas} + \Delta S_{AB}^{res} + \Delta S_{BC}^{gas} + \Delta S_{BC}^{res} + \Delta S_{CA}^{gas} + \Delta S_{CA}^{res}$$
$$= \Delta S_{BC}^{gas} + \Delta S_{BC}^{res}$$

since $\Delta S_{AB}^{gas} = \Delta S_{AB}^{res} = 0$ for the adiabatic process and $\Delta S_{CA}^{gas} = -\Delta S_{CA}^{res}$ for the isothermal process. From this

$$\Delta S_{\circlearrowleft}^{\Sigma} = \Delta S_{BC}^{gas} + \Delta S_{BC}^{res} = \int_{B}^{C} \frac{dQ^{gas}}{T} + \int_{B}^{C} \frac{dQ^{res}}{T^{res}} = \int_{B}^{C} \left(\frac{1}{T} - \frac{1}{T^{res}} \right) dQ^{gas} > 0$$

since $T_D < T_C - T^{res}$ and $Q_{BC}^{gas} > 0$ (heat energy is added to the gas during the isochoric process). You can also show that going clockwise (A $\rightarrow$ C $\rightarrow$ B $\rightarrow$ A) the cycle is also irreversible with $\Delta S_{\circlearrowleft}^{\Sigma} > 0$.

A final note: Work done on a closed system can result in an increase in entropy even when $Q = 0$. A famous example is the experiment performed by Joule in which a force slowly turns paddles in a viscous liquid (see Fig. 6.6). The tank is insulated so no energy enters or leaves the system in the form of heat. The temperature increases due to the work done by the applied force however, unlike adiabatic compression discussed earlier, this is an irreversible process in an inhomogeneous system.

$$* * *$$

This chapter began with a simple example of a rock cycling between a pair of reservoirs. Intuitively we know that this process always leads to energy being transferred from hot reservoir to the cold reservoir. Yet the First Law doesn't forbid the reverse, it only requires that energy be conserved. That's why we need the Second Law, which is the topic of the next chapter.

Chapter 7

Second Law of Thermodynamics

The previous chapter lead us to the conclusion, as stated by Planck, that: "Every process occurring in nature proceeds in the sense in which the sum of the entropies of all bodies taking part in the process is increased. In the (theoretical) limit, i.e., for reversible processes, the sum of the entropies remains unchanged." In this chapter we use this observation to formulate the Second Law of Thermodynamics.

7.1 Gibbs Equation for Closed, Inert Systems

Having introduced entropy in Chapter 5 we can now represent both work and heat in terms of state variables. This allows us to use the First Law to formulate a fundamental thermodynamic identity: the Gibbs equation. There are two commonly used forms of Gibbs equation: taking internal energy as a function of the state variables $(S, V, \mathbf{M})$ is called the *energy representation*; using entropy with $(U, V, \mathbf{M})$ as the state variables is called the *entropy representation*.

Let's start with the energy representation for closed, inert systems; since $\mathbf{M}$ is fixed we write $U(S, V)$. First Law is $dU = dQ + dW$ and on quasistatic paths $dQ = T d_e S$ and $dW = -p dV$. The entropy in these homogeneous systems increases only by heat added so $dS = d_e S$. In summary, we have

$$dU = T dS - p dV \qquad \text{(closed, inert)}.$$

This is the simplest form of the *Gibbs equation*. From multivariate calculus

$$dU(S, V) = \left(\frac{\partial U}{\partial S} \right)_{V, \mathbf{M}} dS + \left(\frac{\partial U}{\partial V} \right)_{S, \mathbf{M}} dV.$$

By comparing this expression with the Gibbs equation

$$T = \left(\frac{\partial U}{\partial S}\right)_{V,M} \qquad \text{and} \qquad p = -\left(\frac{\partial U}{\partial V}\right)_{S,M}.$$

These relations are the fundamental definitions for the equations of state for temperature and pressure. In other words, given $U(S,V)$ we can easily find $T(S,V)$ and $p(S,V)$. Also, from $U(S,V)$ we know $S(U,V)$ so these equations of state can be expressed in the more convenient forms of $T(U,V)$ and $p(T,V)$.

Example 7.1:   The entropy of a given pure system is

$$S(U,V,M) = \gamma M \ln(U + \alpha M) + \delta M \ln(V - \beta M) + f(M)$$

where f is a function of only M; α, β, γ, and δ are constants. Find $T(U,V,M)$ and $p(U,V,M)$.

Solution: For the temperature $T = (\partial U/\partial S)_{V,M}$ so

$$\frac{1}{T} = \left(\frac{\partial S}{\partial U}\right)_{V,M} = \left(\frac{\partial}{\partial U}\right)_{V,M} \{\gamma M \ln(U + \alpha M)\} = \frac{\gamma M}{U + \alpha M}$$

or

$$T(U,V,M) = \frac{U + \alpha M}{\gamma M}.$$

To find the pressure we write

$$p = -\left(\frac{\partial U}{\partial V}\right)_{S,M} = \left(\frac{\partial U}{\partial S}\right)_{V,M}\left(\frac{\partial S}{\partial V}\right)_{U,M}$$

by using the cyclic chain rule (see Section 2.2). From the result for temperature

$$\left(\frac{\partial U}{\partial S}\right)_{V,M} = \left[\left(\frac{\partial S}{\partial U}\right)_{V,M}\right]^{-1} = \frac{U + \alpha M}{\gamma M}.$$

The other derivative is,

$$\left(\frac{\partial S}{\partial V}\right)_{U,M} = \left(\frac{\partial}{\partial V}\right)_{U,M} \{\delta M \ln(V - \beta M)\} = \frac{\delta M}{V - \beta M}$$

so

$$p(U,V,M) = \frac{\delta}{\gamma} \frac{U + \alpha M}{V - \beta M}.$$

Notice that the equation of state for temperature does not contain the constants β and δ so it is not possible to determine $S(U,V,M)$ or $p(U,V,M)$ solely from $T(U,V,M)$. This observation is explained in Section 12.3.

An alternative form of the Gibbs equation is

$$dS = \frac{1}{T}\,dU + \frac{p}{T}\,dV \qquad \text{(closed, inert)},$$

which is used for the entropy formulation $S(U, V)$. Applying the mathematics identity to this differential expression gives

$$\frac{1}{T} = \left(\frac{\partial S}{\partial U}\right)_{V,\mathbf{M}} \qquad \text{and} \qquad \frac{p}{T} = \left(\frac{\partial S}{\partial V}\right)_{U,\mathbf{M}}$$

which provides an alternative expression for pressure

$$p = T \left(\frac{\partial S}{\partial V}\right)_{U,\mathbf{M}}.$$

The energy and entropy formulations are equivalent but one approach is sometimes more convenient than the other.

Example 7.2:   Formulate the Gibbs equation for a closed, inert system in which mechanical work is done by surface tension, γ_{st}, instead of pressure; the surface area of the system is A.

Solution: From Section 3.3 we write the work as $dW = \gamma_{\text{st}}\, dA$ so in the energy representation the Gibbs equation is

$$dU = TdS + \gamma_{\text{st}}\, dA \qquad \text{(Fixed } \mathbf{M}).$$

Note that for this system the temperature and surface tension are defined as

$$T = \left(\frac{\partial U}{\partial S}\right)_{A,\mathbf{M}} \qquad \text{and} \qquad \gamma_{\text{st}} = \left(\frac{\partial U}{\partial A}\right)_{S,\mathbf{M}}.$$

In the entropy representation it is $dS = (1/T)dU - (\gamma_{\text{st}}/T)dA$.

To appreciate the significance of Gibbs equation consider a homogeneous system at thermodynamic equilibrium (e.g., a liter of sea water). We generally need to know both the equations of state for pressure, $p(U, V, \mathbf{M})$, and for temperature, $T(U, V, \mathbf{M})$, to specify the other thermodynamic quantities (c_V, α, κ_T, etc.). And if the composition, $\mathbf{M}$, can change (e.g., add more salt) then we need even more equations of state (see Section 9.1).

On the other hand, if we know $U(S, V, \mathbf{M})$ or $S(U, V, \mathbf{M})$ then *all* other thermodynamic quantities, including all the equations of state, may be derived. In other words, a homogeneous system is entirely represented by these functions with these choices of independent variables. And the key that opens that door is Gibbs equation.

7.2   Second Law – Classical & Modern Forms

This section introduces the Second Law of Thermodynamics as it applies to isolated systems. These systems may be homogeneous, heterogeneous, or continuous; in all cases we assume local equilibrium for the constituents. For example, in the tea cup system of Section 2.3 the hot tea and the cold cup are taken to be homogeneous subsystems separately at thermodynamic equilibrium. The total system is not at equilibrium while $T_{\text{tea}} \neq T_{\text{cup}}$ and energy, in the form of heat, is exchanged between the tea and cup. For this heterogeneous system $S = S_{\text{tea}} + S_{\text{cup}}$ since, being extensive, the entropy of a system is the sum (for heterogenous systems) or the integral (for continuous systems) of entropy over the elements in the system.

For entropy change we write

$$dS = d_{\text{e}}S + d_{\text{i}}S$$

where $d_{\text{e}}S$ is the infintesimal change of entropy due to external sources (e.g., heat energy entering a system) and $d_{\text{i}}S$ is due to internal sources. The change in entropy from state A to state B is independent of the process so by calculating $\Delta S_{\text{AB}} = \int_{\text{A}}^{\text{B}} dS = S_{\text{B}} - S_{\text{A}}$ for a quasi-static process we know it for all processes.

In Classical Thermodynamics the Second Law can be stated as: For isolated systems

$$d_{\text{e}}S = 0 \qquad \text{and} \qquad d_{\text{i}}S \geq 0 \qquad\qquad \text{(Isolated systems)}$$

where the equality holds for reversible processes and at thermodynamic equilibrium. In most textbooks this is simply written as $dS \geq 0$ for isolated systems.

The Modern Thermodynamics version of the Second Law is: For isolated systems

$$\frac{d_{\text{e}}S}{dt} = 0 \qquad \text{and} \qquad \frac{d_{\text{i}}S}{dt} \geq 0 \qquad\qquad \text{(Isolated systems)}$$

where the equality holds for reversible processes and at equilibrium. Modern thermodynamics goes further and shows that the *rate of entropy production* is

$$\frac{d_{\text{i}}S}{dt} = \sum_{\alpha} \mathcal{F}_{\alpha} \mathcal{J}_{\alpha}$$

where $\mathcal{F}_{\alpha}$ is a *thermodynamic force* and $\mathcal{J}_{\alpha}$ is a *thermodynamic flow*.* The summation is over the types of internal flows (e.g., energy, mass, charge) that can occur in a given system. Since $d_{\text{i}}S/dt \geq 0$ this means that $\mathcal{F}_{\alpha}$ and $\mathcal{J}_{\alpha}$ must have the same sign. Despite its name, thermodynamic force is generally *not* a mechanical force, as you'll see when we formulate $\mathcal{F}_{\alpha}$ and $\mathcal{J}_{\alpha}$ for various isolated systems.

---

*The terms "thermodynamic affinity" and "thermodynamic flux" are sometimes used for $\mathcal{F}_{\alpha}$ and $\mathcal{J}_{\alpha}$.

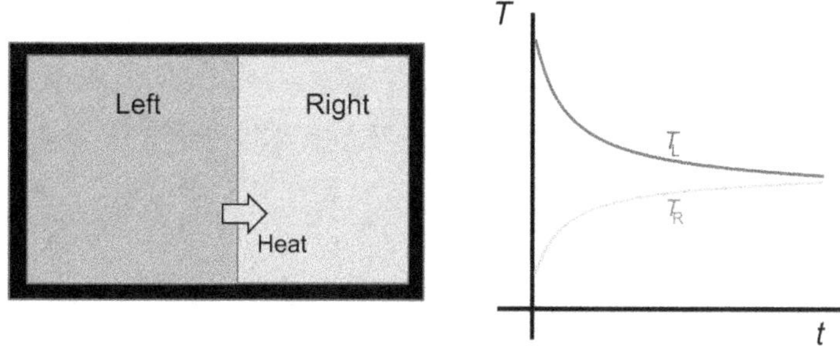

Figure 7.1: (left) Heat conduction in a heterogeneous isolated system; (right) Temperatures $T_L(t)$ and $T_R(t)$ versus time.

7.3 Entropy in Heterogeneous Systems

In this section we'll analyze a heterogeneous system, such as the tea cup example in Section 2.3, to calculate the entropy production due to heat conduction. Consider an isolated system consisting of two homogeneous subsystems, labeled Left (L) and Right (R). The subsystems are inert and closed with a shared conductive, stationary, solid wall between them. The total system is initially not at equilibrium with $T_L > T_R$ but we assume local equilibrium for each subsystem. Over time the flow of heat raises T_R and lowers T_L (see Figure 7.1).

Each subsystem is closed and inert so $\mathbf{M}_L$ and $\mathbf{M}_R$ are fixed; the wall between them is stationary so $dV = 0$ for each side. As such, in the entropy representation the Gibbs equation for each side is simply

$$dS_L = \frac{dU_L}{T_L} \qquad \text{and} \qquad dS_R = \frac{dU_R}{T_R}.$$

By energy conservation $dU_L = -dU_R$ so for the total system

$$dS = \frac{dU_L}{T_L} + \frac{dU_R}{T_R} = \left(\frac{1}{T_R} - \frac{1}{T_L}\right) dU_R$$

since $S = S_L + S_R$. In this isolated system $d_e S = 0$ (no heat enters or leaves) so $dS = d_i S$ and the classical form of the Second Law gives

$$dS = d_i S = \left(\frac{1}{T_R} - \frac{1}{T_L}\right) dU_R \geq 0.$$

In the modern form of the Second Law we write this as a time derivative

$$\frac{d_i S}{dt} = \left(\frac{1}{T_R} - \frac{1}{T_L}\right) \frac{dU_R}{dt} \geq 0.$$

Given that $T_L > T_R$ the term in parenthesis is positive so $dU_R/dt \geq 0$ which means that the energy of the Right subsystem (i.e., the cold side) increases until the temperatures are equal. This transfer of energy is in the form of heat since $dV = 0$ so no work is done on or by either subsystem; by the First Law $dU = dQ$ because $dW = 0$.

Example 7.3: Consider an isolated system consisting of a rock at $T^{\text{rock}} = 300$ K in contact with a hot reservoir at $T^{\text{hot}} = 400$ K (see Fig. 6.1). Find the rate of entropy production for this system given that heat is entering the rock at a rate of 60 J/s.

Solution: From the above, the rate of entropy production is

$$\frac{d_i S}{dt} = \left(\frac{1}{T^{\text{rock}}} - \frac{1}{T^{\text{hot}}} \right) \frac{dU^{\text{rock}}}{dt} = \left(\frac{1}{300 \text{ K}} - \frac{1}{400 \text{ K}} \right) (60 \text{ J/s}) = 0.05 \text{ J/(K s)}.$$

Note that this is for the entire system, that is, both the rock and the reservoir. Over time the rock's temperature increases and the rate of heat flow decreases so $d_i S/dt$ decreases.

The rate of entropy production can be written as

$$\frac{d_i S}{dt} = \mathcal{F}_U \, \mathcal{J}_U$$

and we identify the thermodynamic force and thermodynamic flow for internal energy as

$$\mathcal{F}_U = \frac{1}{T_R} - \frac{1}{T_L} \qquad \text{and} \qquad \mathcal{J}_U = \frac{dU_\rightarrow}{dt}.$$

Here the notation $dU_\rightarrow/dt = dU_R/dt$ is used to remind us that the thermodynamic flow indicates the rate at which the energy is flowing from Left to Right.

As the temperature difference decreases so do the thermodynamic force, $\mathcal{F}_U$, and the thermodynamic flow, $\mathcal{J}_U$. Over time the total entropy increases while the rate of entropy production, $d_i S/dt = \mathcal{F}_U \mathcal{J}_U$, decreases (see Figure 7.2). In general the thermodynamic force, $\mathcal{F}$, can be positive or negative but it always has the same sign as the flow, $\mathcal{J}$, so their product is positive. At thermodynamic equilibrium both $\mathcal{F} = 0$ and $\mathcal{J} = 0$ so $d_i S/dt = 0$ and entropy reaches its maximum value.

The evolution described above implicitly assumes that the T_R increases as energy flows into the Right subsystem and T_L decreases as energy leaves the Left subsystem. In other words, we require that

$$\left(\frac{\partial T}{\partial U} \right)_{V,\mathbf{M}} = \frac{1}{C_V} > 0.$$

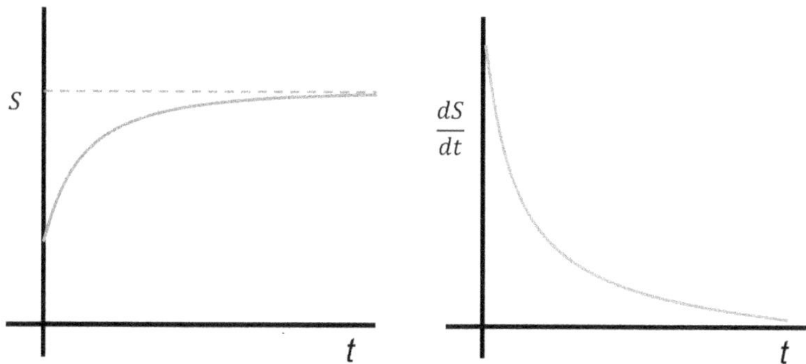

Figure 7.2: Entropy and rate of entropy production in a heterogeneous isolated system.

This establishes the *stability condition* that $C_V > 0$. If heat capacity was negative then the temperature difference, $T_L - T_R$, would increase with time; the Right subsystem would be vampiric in that it would suck all the internal energy out of the Left subsystem.

Example 7.4: Find the entropy production rate for an isolated heterogeneous system consisting of two inert, closed subsystems sharing a mobile wall (i.e., a piston). Take $T_L = T_R = T_0$ but $p_L \neq p_R$.
Solution: From Gibbs equation we have for the left subsystem

$$dS_L = \frac{1}{T_L} dU_L + \frac{p_L}{T_L} dV_L$$

and similarly for the right subsystem. We're assuming equal temperatures so for the total system,

$$dS = dS_L + dS_R = \frac{1}{T_0}(dU_L + dU_R) + \frac{1}{T_0}(p_L dV_L + p_R dV_R) = \frac{p_L - p_R}{T_0} dV_L$$

since for this isolated system $dU_L = -dU_R$ and $dV_L = -dV_R$ (i.e., total energy and volume are conserved). For an isolated system $d_e S = 0$ so $dS = d_i S$; the entropy production rate is

$$\frac{d_i S}{dt} = \left(\frac{p_L - p_R}{T_0} \right) \left(\frac{dV_\rightarrow}{dt} \right) = \mathcal{F}_V \mathcal{J}_V$$

where $dV_\rightarrow = dV_L = -dV_R$. If $p_L > p_R$ then, by the Second Law ($d_i S/dt \geq 0$), we know that $dV_\rightarrow/dt > 0$, which means that the piston moves left to right. Here the stability condition is that $\kappa_T > 0$. Note that the motion of the piston will change the pressures so the temperatures may not remain equal. The system is at *thermal equilibrium* when $T_L = T_R$ and it is at *mechanical equilibrium* when $p_L = p_R$; when both are true then it is at thermodynamic equilibrium.

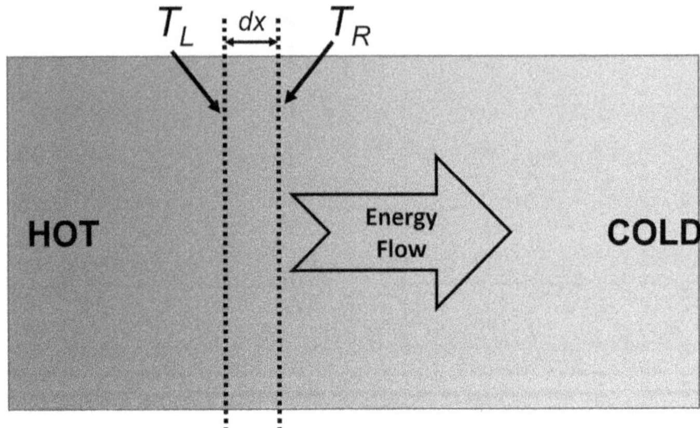

Figure 7.3: Continuous system with a temperature gradient.

7.4   Entropy in Continuous Systems*

The formulation of entropy production for heterogeneous systems may be extended to continuous systems by replacing differences with derivatives. To see how this works consider the one-dimensional system illustrated in Figure 7.3. The full system is out of equilibrium but we assume *local equilibrium* for a thin slab of thickness Δx, allowing us to define a local temperature, $T(x)$, that varies but only in the x direction.

We start with the rate of entropy production

$$\frac{d_i S}{dt} = \mathcal{F}_U \mathcal{J}_U = \left(\frac{1}{T_R} - \frac{1}{T_L} \right) \left(\frac{dU_\rightarrow}{dt} \right).$$

For small Δx the thermodynamic force may written as[†]

$$\mathcal{F}_U = \frac{1}{T_R} - \frac{1}{T_L} \approx -\frac{T_R - T_L}{T^2} = -\frac{1}{T^2}\frac{T_R - T_L}{\Delta x}\Delta x \approx -\frac{1}{T^2}\frac{dT}{dx}\Delta x$$

where $T = \frac{1}{2}(T_L + T_R)$ is the average temperature in the slab. For continuous systems it makes sense to redefine the thermodynamic force and flow as

$$\begin{aligned}
\bar{\mathcal{F}}_U &= \frac{1}{\Delta x}\mathcal{F}_U = -\frac{1}{T^2}\frac{dT}{dx} \\
\bar{\mathcal{J}}_U &= \frac{1}{A_s}\mathcal{J}_U = \frac{1}{A_s}\frac{dU_\rightarrow}{dt}
\end{aligned}$$

[†]For this continuous system $T(x, t)$ so one should write $\partial T/\partial x$ instead of dT/dx. The reason we don't is only to avoid confusion with other uses of partial derivatives (e.g., $(\partial T/\partial V)_{p,\mathbf{M}}$)

where A_s and $V_\mathrm{s} = A_\mathrm{s}\Delta x$ are the cross-section surface area and volume of the thin slab. Notice that $\bar{\mathcal{J}}_U$ is the heat flux, that is, the rate at which heat energy is flowing per unit area. For continuous systems we define

$$\varsigma(x) = \frac{1}{V_\mathrm{s}}\frac{d_\mathrm{i}S}{dt} = \frac{1}{V_\mathrm{s}}\mathcal{F}_U\mathcal{J}_U = \bar{\mathcal{F}}_U\bar{\mathcal{J}}_U$$

as the *local rate of entropy production*.

Entropy can also change due to an external source, specifically in this case if there's a difference between the heat entering and leaving the slab. This is

$$\begin{aligned}
\frac{d_\mathrm{e}S}{dt} &= \left\{\frac{1}{T}\frac{dU_\rightarrow}{dt}\right\}_\mathrm{L} - \left\{\frac{1}{T}\frac{dU_\rightarrow}{dt}\right\}_\mathrm{R} = \frac{\mathcal{J}_{U,\mathrm{L}}}{T_\mathrm{L}} - \frac{\mathcal{J}_{U,\mathrm{R}}}{T_\mathrm{R}} \\
&= -V_\mathrm{s}\frac{(\bar{\mathcal{J}}_{U,\mathrm{R}}/T_\mathrm{R}) - (\bar{\mathcal{J}}_{U,\mathrm{L}}/T_\mathrm{L})}{\Delta x} \approx -V_\mathrm{s}\frac{d}{dx}\left(\frac{\bar{\mathcal{J}}_U}{T}\right).
\end{aligned}$$

Defining the entropy density, $s(x) = S(x)/V_\mathrm{s}$, combining the above gives

$$\frac{ds}{dt} = \frac{d_\mathrm{i}s}{dt} + \frac{d_\mathrm{e}s}{dt} = \varsigma - \frac{d}{dx}\frac{\bar{\mathcal{J}}_U}{T}.$$

which is the *continuity equation* for entropy. A system is at a non-equilibrium steady state when $\varsigma = -d_\mathrm{e}s/dt$, that is, if the internal entropy production is balanced by the loss of entropy due to heat flow (see example below).

Example 7.5: The ends of a metal bar (length ℓ, cross-section $A_\mathrm{s} = 1$ cm^2) are held steady at temperatures $T_\mathrm{L} = 400$ K and $T_\mathrm{R} = 300$ K resulting in a constant heat flux of 6 kilowatts/cm^2. The sides of the bar are thermally insulated (i.e., adiabatic walls). Calculate the amount of entropy (in J/K) produced per second inside the metal bar and show that this equals the rate at which entropy leaves from the ends.

Solution: The rate of internal entropy production is

$$\begin{aligned}
\frac{d_\mathrm{i}S}{dt} &= \int \varsigma\, dV = A_\mathrm{s}\int_0^\ell \bar{\mathcal{F}}_U\bar{\mathcal{J}}_U\, dx = A_\mathrm{s}\int_0^\ell \left(-\frac{1}{T^2}\frac{dT}{dx}\right)\bar{\mathcal{J}}_U\, dx \\
&= -A_\mathrm{s}\bar{\mathcal{J}}_U\int_{T_\mathrm{L}}^{T_\mathrm{R}}\frac{dT}{T^2} = A_\mathrm{s}\bar{\mathcal{J}}_U\left(\frac{1}{T_\mathrm{R}} - \frac{1}{T_\mathrm{L}}\right) \\
&= (1.0\text{ cm}^2)(6000\text{ J/(s cm}^2)) \left(\frac{1}{300\text{ K}} - \frac{1}{400\text{ K}}\right) = 5\text{ J/(s K)}.
\end{aligned}$$

The heat added per second at the bar's left end is $\mathcal{J}_\mathrm{L} = 6000$ J/s so the entropy added per second at this end is $(d_\mathrm{e}S/dt)_\mathrm{L} = \mathcal{J}_\mathrm{L}/T_\mathrm{L}$. Similarly, for the right end $\mathcal{J}_\mathrm{R} = -6000$ J and $(d_\mathrm{e}S/dt)_\mathrm{R} = \mathcal{J}_\mathrm{R}/T_\mathrm{R}$. The total entropy added per second at the ends is

$$\frac{d_\mathrm{e}S}{dt} = \left(\frac{d_\mathrm{e}S}{dt}\right)_\mathrm{L} + \left(\frac{d_\mathrm{e}S}{dt}\right)_\mathrm{R} = \frac{6000\text{ J/s}}{400\text{ K}} + \frac{-6000\text{ J/s}}{300\text{ K}} = -5\text{ J/(s K)}.$$

so the internal entropy production is balanced by entropy removal at the boundaries. Finally, note that while the system is steady (i.e., temperature is not changing with time) it is *not* at thermodynamic equilibrium since $\varsigma \neq 0$. The system is said to be at a non-equilibrium steady state.

Be careful not to equate $dU_{\rightarrow}/dt$ with dU/dt, the rate at which the internal energy is changing. The latter is given by

$$\frac{dU}{dt} = \mathcal{J}_{U,\mathrm{L}} - \mathcal{J}_{U,\mathrm{R}} = -V_{\mathrm{s}}\frac{\bar{\mathcal{J}}_{U,\mathrm{R}} - \bar{\mathcal{J}}_{U,\mathrm{L}}}{\Delta x}$$

so for $\Delta x \rightarrow 0$

$$\frac{du}{dt} = -\frac{d}{dx}\bar{\mathcal{J}}_U$$

where $u = U/V_{\mathrm{s}}$ is the internal energy density. In the example above energy flows from left to right ($\bar{\mathcal{J}}_U > 0$) yet the system is at a steady state ($du/dt = 0$) so at any given point the same amount enters and leaves ($d\bar{\mathcal{J}}_U/dx = 0$).

For three-dimensional systems these results generalize to give

$$\frac{d}{dt}s(\mathbf{x}) = \varsigma - \nabla \cdot \left(\frac{\bar{\boldsymbol{\mathcal{J}}}_U}{T}\right)$$

where

$$\varsigma(\mathbf{x}) = \bar{\boldsymbol{\mathcal{F}}}_U \cdot \bar{\boldsymbol{\mathcal{J}}}_U \qquad \text{with} \qquad \bar{\boldsymbol{\mathcal{F}}}_U = -\frac{1}{T^2}\nabla T = \nabla(1/T)$$

and ∇ is the gradient operator. The energy flux is given by the continuity equation for internal energy

$$\frac{d}{dt}u(\mathbf{x}) = -\nabla \cdot \bar{\boldsymbol{\mathcal{J}}}_U$$

which is simply another way of writing energy conservation.

Finally, as we'll see in the Chapters 9 and 10, metabolic processes in living systems continuously produce entropy due to chemical reactions. As illustrated in the example above, a system can sustain a non-equilibrium steady state by expelling internally produced entropy. Living creatures, such as a cat, take in nutrients and expel wastes, balancing the internal entropy production with a net outflow of entropy. Schrödinger referred to this outflow as "negative entropy", a thermodynamic requirement for all living organisms.

$$* \ * \ *$$

The Second Law of Thermodynamics is considered by many as one of the most important principles in all of science.

It is a remarkable fact that the second law of thermodynamics has played in the history of science a fundamental role far beyond its original scope. Suffice it to mention Boltzmann's work on kinetic theory, Planck's discovery of quantum theory or Einstein's theory of spontaneous emission, which were all based on the second law of thermodynamics. – Ilya Prigogine (Nobel lecture, 1977)

In the next few chapters we'll explore additional facets of the Second Law as we apply it in a variety of scenarios.

Chapter 8

Thermodynamic Cycles

Naively, the Second Law appears to say that heat always flows from hot to cold yet we know that refrigerators can exist. This chapter shows how a thermodynamic cycle can use work to move energy repeatedly from a cold reservoir to a hot reservoir. A thermodynamic cycle can also do the reverse (i.e., use heat to do work) and that's our first topic.

8.1 Carnot Cycle

Heat Engines

Heat engines are described by a sequence of processes forming a closed cycle. For example, Figure 8.1 illustrates the Otto engine cycle, which consists of four quasi-static branches: two adiabatic paths and two isochoric (constant volume) paths. The net work done by this engine over one cycle is

$$W_{\circlearrowleft} = \oint dW = \int_A^B p(V)dV + \int_B^C p(V)dV + \int_C^D p(V)dV + \int_D^A p(V)dV.$$

We're interested in the work done by the engine, $\mathcal{W} = -W$, so we'll write the First Law as $\Delta U = Q - \mathcal{W}$. In general, a cycle returns to its initial state $\Delta U_{\circlearrowleft} = 0$ so by the First Law

$$\oint dQ = \oint d\mathcal{W} \qquad \text{or} \qquad Q_{\circlearrowleft} = \mathcal{W}_{\circlearrowleft}$$

where $Q_{\circlearrowleft}$ is the net heat added to the system. For an engine $Q_{\circlearrowleft} > 0$ so heat energy is added to the system with $\mathcal{W}_{\circlearrowleft} > 0$. Refrigerators and heat pumps are the same as heat engines except that $Q_{\circlearrowleft} < 0$ and $\mathcal{W}_{\circlearrowleft} < 0$. For refrigerators we care about where the heat comes from and for heat pumps we care about where the heat goes (e.g., a heat pump warms your home on a cold day).

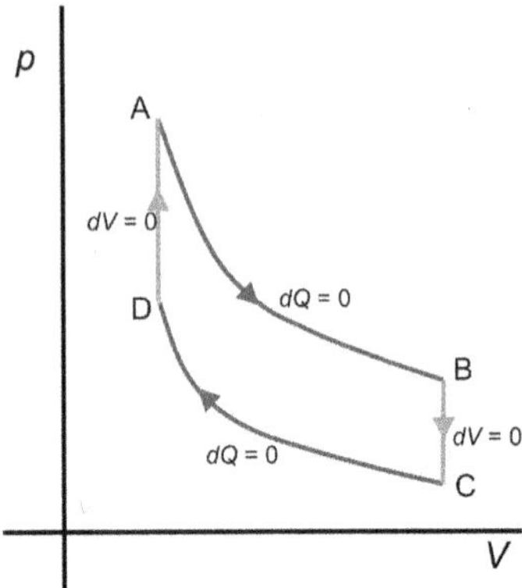

Figure 8.1: Otto engine cycle in $p - V$ diagram. The branches A $\rightarrow$ B and C $\rightarrow$ D are adiabatic paths; the branches B $\rightarrow$ C and D $\rightarrow$ A are isochoric paths.

The *Carnot cycle* is an important, special type of cycle in which the entire path is reversible. The simplest version of a Carnot cycle consists of a cycle pieced together from a pair of reversible isothermal paths and a pair of reversible adiabatic paths (see Figure 8.2). The Carnot cycle looks complicated in the $p-V$ diagram; for an ideal gas it's a funky, curved trapezoid. It's much simpler in the $T - S$ diagram; the Carnot cycle is just a rectangle. In fact, it's a rectangle *independent* of the properties of the system (i.e., it doesn't have to be an ideal gas). The two isothermal branches are horizontal lines and the two adiabatic branches are vertical lines.

In the $T - S$ diagram on the branch of the cycle that goes from left to right, heat energy is being added to the system by the hot reservoir

$$Q_{\text{hot}} = \int_A^B T_{\text{hot}} \, d_e S = T_{\text{hot}} \int_A^B d_e S = T_{\text{hot}}(S_B - S_A) = T_{\text{hot}} \Delta S_{AB} > 0$$

which is the area under this branch of the cycle in the $T - S$ diagram. Since an engine is a cycle, there is at least one branch during which the entropy decreases. For the Carnot engine this occurs on the branch when the system is connected to the cold reservoir.* On the lower isothermal branch

$$Q_{\text{cold}} = \int_C^D T_{\text{cold}} \, d_e S = T_{\text{cold}}(S_D - S_C) = -T_{\text{cold}} \Delta S_{AB} < 0$$

since $S_A = S_D$ and $S_B = S_C$. Finally, the area enclosed in the $T - S$ diagram

---

*For a refrigerator the cycle goes counter-clockwise so these are switched ($Q_{\text{cold}} > 0$ and $Q_{\text{hot}} < 0$) so heat energy is drawn from the cold reservoir and dumped into the hot reservoir.

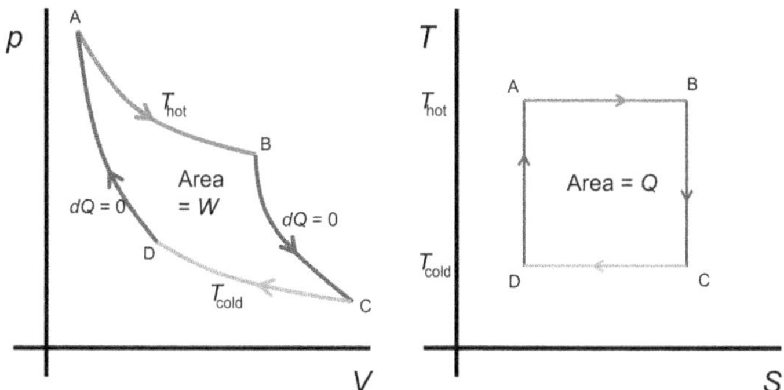

Figure 8.2: Carnot engine cycle in $p-V$ diagram (left) and $T-S$ diagram (right). The branches $A \rightarrow B$ and $C \rightarrow D$ are isothermal paths; the branches $B \rightarrow C$ and $D \rightarrow A$ are adiabatic paths. The enclosed areas equal $\mathcal{W}_\circlearrowright$ (left) and $Q_\circlearrowright$ (right).

is the net heat added to the system so

$$Q_\circlearrowright = \oint dQ = \oint T d_e S = (T_{\text{hot}} - T_{\text{cold}})\Delta S_{\text{AB}} > 0$$

and the work done by the system is $\mathcal{W}_\circlearrowright = Q_\circlearrowright$.

Engine Efficiency

Since $Q_\circlearrowright = \mathcal{W}_\circlearrowright$, does that mean that a Carnot engine is 100 percent efficient? No, in fact we get this result for *all* engines; it's just a consequence of energy conservation (i.e., First Law). To explain, let me put it in economic terms: Say you're in the business of manufacturing widgets and this month you had a gross earnings of $1000. If your expenses are $900 then your profit is $100. As an economic engine $Q_\circlearrowright = \$1000 - \$900$ and $\mathcal{W}_\circlearrowright = \$100$ so the "business efficiency" is only 10 percent since $\$100/\$1000 = 0.1$.

The engine efficiency is defined as

$$\eta = \frac{\mathcal{W}_\circlearrowright}{Q_{\text{hot}}}$$

which is the fraction of the heat energy taken in by the engine that gets converted to work. This may also be written as

$$\eta = \frac{Q_{\text{hot}} + Q_{\text{cold}}}{Q_{\text{hot}}} = 1 + \frac{Q_{\text{cold}}}{Q_{\text{hot}}} = 1 - \frac{|Q_{\text{cold}}|}{|Q_{\text{hot}}|} \qquad \text{(All Engines)}$$

since by First Law $\mathcal{W}_\circlearrowright = Q_{\text{hot}} + Q_{\text{cold}}$.

The efficiency of the Carnot engine is

$$\eta = 1 + \frac{Q_{\text{cold}}}{Q_{\text{hot}}} = 1 + \frac{T_{\text{cold}}(S_{\text{A}} - S_{\text{B}})}{T_{\text{hot}}(S_{\text{B}} - S_{\text{A}})} = 1 - \frac{T_{\text{cold}}}{T_{\text{hot}}} \qquad \text{(Carnot Engine)},$$

which is a simple relation involving the ratio of temperatures. William Thomson (Lord Kelvin) established the absolute temperature scale using the fact that the efficiency of a Carnot engine only depends on the temperatures of the reservoirs and not on the properties of the materials.

Example 8.1: A Carnot refrigerator is a reversible cycle that's identical to the Carnot engine but goes counter-clockwise in the $T - S$ diagram. The coefficient of performance of a refrigerator is

$$\mathcal{K} = \frac{Q_{\text{cold}}}{W_{\circlearrowleft}}$$

where Q_{cold} is the heat removed from the cold reservoir and $W_{\circlearrowleft}$ is the work done on the refrigerator in the cycle. This coefficient is typically around $\mathcal{K} \approx 3$ for modern air conditioners.

Solution: For a Carnot refrigerator both isothermal paths are reversible so

$$Q_{\text{cold}} = T_{\text{cold}}\Delta S_{\text{DC}} \qquad \text{and} \qquad Q_{\text{hot}} = T_{\text{hot}}\Delta S_{\text{BA}} = -T_{\text{hot}}\Delta S_{\text{DC}}$$

For the cycle $\Delta U = 0$ so by First Law $W_{\circlearrowleft} + Q_{\circlearrowleft} = 0$ thus $W_{\circlearrowleft} = -Q_{\circlearrowleft} = -(Q_{\text{cold}} + Q_{\text{hot}}) = (T_{\text{hot}} - T_{\text{cold}})\Delta S_{\text{DC}}$. Finally the coefficient of performance is

$$\mathcal{K} = \frac{Q_{\text{cold}}}{W_{\circlearrowleft}} = \frac{T_{\text{cold}}\Delta S_{\text{DC}}}{(T_{\text{hot}} - T_{\text{cold}})\Delta S_{\text{DC}}} = \frac{T_{\text{cold}}}{T_{\text{hot}} - T_{\text{cold}}}.$$

For $T_{\text{hot}} = 40°C$, $T_{\text{cold}} = 20°C$ this gives $\mathcal{K} = 14.7$.

Entropy and the Carnot cycle

Back in Chapter 5 entropy was introduced by the relation $d_{\text{e}}S = dQ/T$. In this chapter we used this relation in the analysis of the Carnot cycle. Historically the concept of entropy came about from the observation that for a Carnot engine

$$\frac{Q_{\text{cold}}}{Q_{\text{hot}}} = -\frac{T_{\text{cold}}}{T_{\text{hot}}}$$

so for this reversible cycle,

$$\frac{Q_{\text{hot}}}{T_{\text{hot}}} + \frac{Q_{\text{cold}}}{T_{\text{cold}}} = 0.$$

This led Clausius to realize that there was something special about dQ/T, specifically that,

$$\oint \frac{dQ}{T} = 0$$

indicating that the integrand was a state variable, similarly to,

$$-\oint \frac{dW}{p} = \oint dV = 0.$$

This train of thought resulted in the definition of entropy as $dS = dQ/T$ when $d_{\mathrm{i}}S = 0$.

8.2   Imperfect Engines*

Consider an imperfect engine, similar to the Carnot engine, with the following difference: On the branch $A \to B$ in which the system is in contact with the hot reservoir we quickly increase the volume and go by a non-equilibrium process from state A to B (see Figure 8.3). The change in entropy going from state A to B is

$$\Delta S_{\mathrm{AB}} = \int_{\mathrm{A}}^{\mathrm{B}} dS = \int_{\mathrm{A}}^{\mathrm{B}} d_{\mathrm{e}}S + \int_{\mathrm{A}}^{\mathrm{B}} d_{\mathrm{i}}S$$

For a continuous system we can, in principle, calculate the local entropy production (see Section 7.4) however we'll simply write the right hand side as

$$\Delta S_{\mathrm{AB}} = \frac{Q_{\mathrm{hot}}}{T_{\mathrm{hot}}} + \Delta S^{\mathrm{irr}}$$

with $\Delta S^{\mathrm{irr}} \geq 0$ being the change in entropy *not* due to heat from the reservoir. In other words,

$$Q_{\mathrm{hot}} = T_{\mathrm{hot}}(\Delta S_{\mathrm{AB}} - \Delta S^{\mathrm{irr}}).$$

We take the other three branches of the cycle to be reversible adiabatic and isothermal processes, same as in the Carnot cycle.

The First Law still gives us that the work done by the imperfect engine, $\mathcal{W}_{\circlearrowright}$, equals the net heat added, $Q_{\circlearrowright} = Q_{\mathrm{hot}} + Q_{\mathrm{cold}}$. This imperfect engine does less work than a similar Carnot engine, specifically,

$$\mathcal{W}_{\circlearrowright}^{\mathrm{lost}} \equiv \mathcal{W}_{\circlearrowright}^{\mathrm{Carnot}} - \mathcal{W}_{\circlearrowright} = Q_{\mathrm{hot}}^{\mathrm{Carnot}} - Q_{\mathrm{hot}} = T_{\mathrm{hot}}\Delta S^{\mathrm{irr}}.$$

Clausius referred to $T_{\mathrm{hot}}\Delta S^{\mathrm{irr}}$ as the 'uncompensated heat' and $\mathcal{W}_{\circlearrowright}^{\mathrm{lost}}$ is known as the 'exergy loss.'[†]

The efficiency of this imperfect engine is less than that of the Carnot engine. Specifically, since $\eta = 1 - \left| \frac{Q_{\mathrm{cold}}}{Q_{\mathrm{hot}}} \right|$ the efficiency is

$$\eta = 1 - \left| \frac{T_{\mathrm{cold}}\Delta S_{\mathrm{CD}}}{T_{\mathrm{hot}}(\Delta S_{\mathrm{AB}} - \Delta S^{\mathrm{irr}})} \right| = 1 - \frac{T_{\mathrm{cold}}}{T_{\mathrm{hot}}(1 - \Delta S^{\mathrm{irr}}/\Delta S_{\mathrm{AB}})} = 1 - \frac{T_{\mathrm{cold}}}{T_{\mathrm{hot}}^{\mathrm{irr}}}$$

[†]A. Bejan, *Entropy Generation Minimization*, CRC Press (2013).

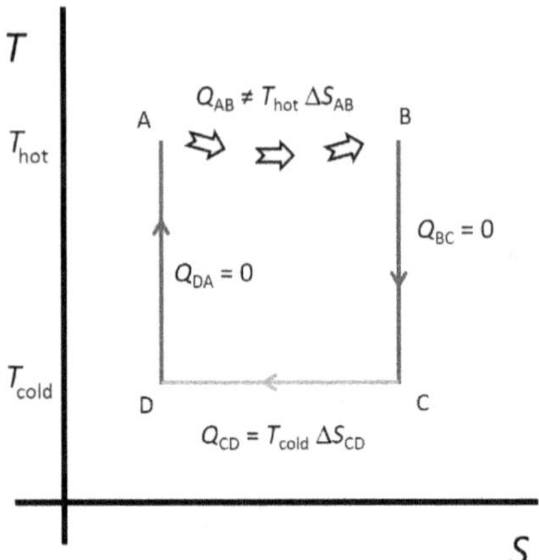

Figure 8.3: $T - S$ diagram for a heat engine that is similar to a Carnot cycle but with an irreversible process from state A to state B.

since $\Delta S_{\mathrm{AB}} = -\Delta S_{\mathrm{CD}}$. In effect, the irreversibility results in the engine operating like a Carnot engine but with the hot reservoir having a reduced temperature of

$$T_{\mathrm{hot}}^{\mathrm{irr}} = T_{\mathrm{hot}} \left(1 - \frac{\Delta S^{\mathrm{irr}}}{\Delta S_{\mathrm{AB}}} \right)$$

so $\eta = 0$ if $T_{\mathrm{hot}}^{\mathrm{irr}} = T_{\mathrm{cold}}$.

Example 8.2: Consider an imperfect Carnot engine in which the path from the hot to cold temperature is not adiabatic (see Fig. 8.4). On this branch the process is irreversible with $\Delta S_{\mathrm{BC}}^{\mathrm{irr}} > 0$. The rest of the cycle's processes are reversible. Find the efficiency of this imperfect Carnot engine in terms of T_{hot}, T_{cold}, ΔS_{AB}, and $\Delta S_{\mathrm{BC}}^{\mathrm{irr}}$.

Solution: The heat from the hot reservoir is $Q_{\mathrm{hot}} = T_{\mathrm{hot}} \Delta S_{\mathrm{AB}} > 0$ and the heat from the cold reservoir is

$$Q_{\mathrm{cold}} = T_{\mathrm{cold}} \Delta S_{\mathrm{CD}} = -T_{\mathrm{cold}}(\Delta S_{\mathrm{AB}} + \Delta S_{\mathrm{BC}}^{\mathrm{irr}}) < 0$$

Since $\Delta U_{\circlearrowleft} = 0$, by First Law $W_{\circlearrowleft} = -Q_{\circlearrowleft}$. Since $W_{\circlearrowleft} = -\mathcal{W}_{\circlearrowleft}$ the efficiency is

$$\eta = \frac{\mathcal{W}_{\circlearrowleft}}{Q_{\mathrm{hot}}} = \frac{Q_{\mathrm{hot}} + Q_{\mathrm{cold}}}{Q_{\mathrm{hot}}} = \frac{T_{\mathrm{hot}} \Delta S_{\mathrm{AB}} - T_{\mathrm{cold}}(\Delta S_{\mathrm{AB}} + \Delta S_{\mathrm{BC}}^{\mathrm{irr}})}{T_{\mathrm{hot}} \Delta S_{\mathrm{AB}}}$$

so

$$\eta = 1 - \frac{T_{\mathrm{cold}}}{T_{\mathrm{hot}}} \left(1 + \frac{\Delta S_{\mathrm{BC}}^{\mathrm{irr}}}{\Delta S_{\mathrm{AB}}} \right)$$

Notice that the efficiency goes to zero when $\Delta S_{\mathrm{BC}}^{\mathrm{irr}}$ is large enough that all of the energy drawn from the hot reservoir is returned to the cold reservoir, in which case $\mathcal{W}_{\circlearrowleft} = 0$.

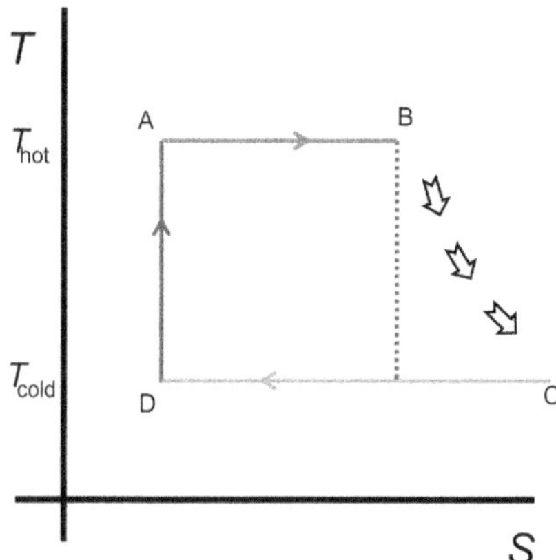

Figure 8.4: $T - S$ diagram for a heat engine that is similar to a Carnot cycle but with an irreversible, non-equilibrium process from state B to state C (arrows) replacing the reversible, adiabatic path (dotted line).

Example 8.3: Consider an imperfect Carnot refrigerator in which the isothermal process at the cold temperature (i.e., from point D to C) is irreversible while the rest of the cycle's processes are reversible. Call ΔS_{DC} the change of entropy if this isothermal process were reversible and ΔS^{irr} the total amount of irreversible entropy that is produced between points D and C. Find the coefficient of performance, $\mathcal{K} = Q_{\text{cold}}/W_{\circlearrowright}$, in terms of T_{hot}, T_{cold}, ΔS^{irr}, and ΔS_{DC}.

Solution: For the imperfect refrigerator

$$Q_{\text{cold}} = T_{\text{cold}}(\Delta S_{\text{DC}} - \Delta S^{\text{irr}}) \qquad \text{and} \qquad Q_{\text{hot}} = T_{\text{hot}}\Delta S_{\text{BA}} = -T_{\text{hot}}\Delta S_{\text{DC}}.$$

For the cycle $\Delta U_{\circlearrowright} = 0$ so by First Law $W_{\circlearrowright} + Q_{\circlearrowright} = 0$ thus

$$W_{\circlearrowright} = -Q_{\circlearrowright} = -(Q_{\text{cold}} + Q_{\text{hot}}) = T_{\text{hot}}\Delta S_{\text{DC}} - T_{\text{cold}}(\Delta S_{\text{DC}} - \Delta S^{\text{irr}}).$$

Finally the coefficient of performance is

$$\mathcal{K} = \frac{Q_{\text{cold}}}{W_{\circlearrowright}} = \frac{T_{\text{cold}}(\Delta S_{\text{DC}} - \Delta S^{\text{irr}})}{T_{\text{hot}}\Delta S_{\text{DC}} - T_{\text{cold}}(\Delta S_{\text{DC}} - \Delta S^{\text{irr}})} = \frac{T_{\text{cold}}^{\text{irr}}}{T_{\text{hot}} - T_{\text{cold}}^{\text{irr}}}$$

where the effective cold temperature is

$$T_{\text{cold}}^{\text{irr}} = T_{\text{cold}}\left(1 - \frac{\Delta S^{\text{irr}}}{\Delta S_{\text{DC}}}\right).$$

If $\Delta S^{\text{irr}} = \Delta S_{\text{DC}}$ then the coefficient of performance $\mathcal{K} = 0$.

$$* * *$$

Up to now we've focused on closed, inert systems but in the next chapter we'll consider systems with transformations (e.g., melting ice) and open systems (e.g., breathing cat). For such systems we have another equation of state, the chemical potential, another thermodynamic force, the affinity, and fluxes of mass. Fortunately these new elements simply augment the First and Second Law of Thermodynamics without changing these foundations.

Chapter 9

Open and Reactive Systems

Reactive systems are interesting since transformations allow for changes in composition. Similarly, composition can change in open systems due to fluxes. In this chapter we consider such systems, starting with the introduction of a new equation of state: chemical potential.

9.1 Chemical Potential

So far we've focused on systems where the mass and composition were fixed. Now we'll consider situations where either the total mass, M, changes or there are changes in the composition, $\mathbf{M} = \{M_1, M_2, \ldots, M_K\}$, or both. Two common situations in which this occurs are:

- **Open Systems** — Mass can enter or leave the system through porous walls.

- **Reactive Systems** — The composition can change due to transformations, such as changes of phase or chemical reactions.

For example, an open, pure system can have $dM \neq 0$ while in a closed, reactive system we can have $dM_k \neq 0$ but $\sum_k dM_k = 0$ by mass conservation. We'll distinguish between changes due to external processes, $d_e M_k$, and those due to internal processes, $d_i M_k$, with $dM_k = d_e M_k + d_i M_k$.

The generalization of the Gibbs equation for open and reactive systems is

$$dU = TdS - pdV + \sum_{k=1}^{K} \mu_k dM_k$$

or

$$dU = TdS - pdV + \boldsymbol{\mu} \cdot d\mathbf{M}$$

where μ_k is the *chemical potential* for species k and $\boldsymbol{\mu} = \{\mu_1, \mu_2, \ldots, \mu_K\}$. Chemical potential is an intensive state variable defined as an energy per unit mass (but $\mu \neq U/M$). An alternative definition is $\hat{\mu}_k = m_k N_A \mu_k$, which is an energy per mole; note that $\mu_k dM_k = \hat{\mu}_k dN_k$.*

For pure systems ($K = 1$), the Gibbs equation simplifies to

$$dU = TdS - pdV + \mu dM \qquad \text{(pure system)}.$$

By writing $U(S, V, M)$ we have the mathematics identity from multivariate calculus

$$dU(S, V, M) = \left(\frac{\partial U}{\partial S}\right)_{V,M} dS + \left(\frac{\partial U}{\partial V}\right)_{S,M} dV + \left(\frac{\partial U}{\partial M}\right)_{S,V} dM$$

Comparing this expression with the Gibbs equation we see that chemical potential is defined as

$$\mu = \left(\frac{\partial U}{\partial M}\right)_{S,V}.$$

This defines $\mu(S, V, M)$ as another equation of state along with pressure and temperature.

Using the entropy representation of the Gibbs equation

$$dS = \frac{1}{T}dU + \frac{p}{T}dV - \frac{\mu}{T}dM \qquad \text{(pure system)}$$

and applying the same math identity gives an alternative definition

$$\mu = -T\left(\frac{\partial S}{\partial M}\right)_{U,V}.$$

The term "chemical" in the name is somewhat misleading because even an inert system has a non-zero chemical potential.

Example 9.1: Find the chemical potential for a pure ideal gas with constant c_V.
Solution: We start with the result from Section 5.2,

$$S(T, V, M) = M\left[s_0 + \frac{k_B}{m}\ln(V/M) + c_V \ln(T)\right]$$

and the relation for an ideal gas, $U(T, M) = c_V M T$, so

$$S(U, V, M) = M\left[s_0 + \frac{k_B}{m}\ln(V/M) + c_V \ln(U/(c_V M))\right].$$

We can now find the chemical potential by simply taking the derivative

$$\mu = -T\left(\frac{\partial S}{\partial M}\right)_{U,V} = -T\left(\frac{1}{M}S - \frac{k_B}{m} - c_V\right) = -\frac{1}{M}TS + c_p T.$$

System	Gibbs equation	Constraints
Open, reactive	$dU = TdS - pdV + \boldsymbol{\mu} \cdot d\mathbf{M}$	None
Open, inert	$dU = TdS - pdV + \boldsymbol{\mu} \cdot d_e\mathbf{M}$	$d_i\mathbf{M} = 0$
Open, pure	$dU = TdS - pdV + \mu d_e M$	$K = 1, d_i M = 0$
Closed, reactive	$dU = TdS - pdV + \boldsymbol{\mu} \cdot d_i\mathbf{M}$	$d_e\mathbf{M} = 0$
Closed, inert	$dU = TdS - pdV$	$d\mathbf{M} = 0$
Isolated, reactive	$0 = Td_iS + \boldsymbol{\mu} \cdot d_i\mathbf{M}$	$dU = d_eS = dV = d_e\mathbf{M} = 0$
Isolated, inert	$0 = Td_iS$	$dU = d_eS = dV = d\mathbf{M} = 0$

Table 9.1: Gibbs equation for various homogeneous systems.

From this expression you can verify that $\mu \neq U/M$. We'll derive more convenient expressions for chemical potential in the later chapters.

The chemical potential is not very interesting for pure systems since $\mu(T, p)$ when there is only one species (see the Gibbs-Duhem equation in Section 13.1). In other words, chemical potential is a function of temperature and pressure so it is not an independent state variable in pure systems. On the other hand, the chemical potential is very useful when there are transformations, such as changes of phase or chemical reactions, in a mixture (see the next section). From the energy and entropy representations we find

$$\mu_k = \left(\frac{\partial U}{\partial M_k} \right)_{S,V,M_{j\neq k}} = -T \left(\frac{\partial S}{\partial M_k} \right)_{U,V,M_{j\neq k}}$$

where the subscript notation $M_{j\neq k}$ means hold M_j fixed for all species *except* species k.

9.2 Material Transformations

Transformations in Thermodynamics

In thermodynamics we're interested in many kinds of transformations. For example, the transformation between solid ice and liquid water

$$H_2O \; (s) \rightleftharpoons H_2O \; (\ell).$$

Chemical reactions are another type of transformation, take

$$2H_2 \; (g) + O_2 \; (g) \rightleftharpoons 2\, H_2O \; (\ell)$$

Another example is the absorption of a photon by a helium atom, which may be written as

$$He + \gamma \rightleftharpoons He^*$$

---

*Here μ is the specific chemical potential and $\bar{\mu}$ is the molar chemical potential.

where asterisk designates the excited state. Similarly, in semiconductors we have the generation of electron-hole pairs by photon absorbtion. In nuclear physics we have,

$$\mathrm{^{6}_{3}Li} + \mathrm{^{2}_{1}H} \rightleftharpoons 2\,\mathrm{^{4}_{2}He} \qquad \text{and} \qquad \mathrm{^{238}_{92}U} \rightleftharpoons \mathrm{^{234}_{90}Th} + \mathrm{^{4}_{2}He},$$

which are examples of a fusion reaction and a radioactive decay.

We'll formulate transformations such as the above in generic form. For example, the phase transition between solid ice and liquid water may be written as

$$\mathbb{X} \rightleftharpoons \mathbb{Y}$$

with "$\mathbb{X}$" being a molecule in the solid phase and "$\mathbb{Y}$" being one in the liquid phase. Similarly, we'll write

$$2\mathbb{A} + \mathbb{B} \rightleftharpoons 2\mathbb{C}$$

for a chemical reaction in which two molecules of "$\mathbb{A}$" and one molecule of "$\mathbb{B}$" transform into two molecules of "$\mathbb{C}$." Using generic expressions highlights that in thermodynamics these various material transformations are treated in a similar fashion.

For a transformation of the form

$$a\,\mathbb{A} + b\,\mathbb{B} + \ldots \rightleftharpoons a'\mathbb{A} + b'\mathbb{B} + \ldots$$

we define the *stoichiometric coefficients*,

$$\nu_{\mathbb{A}} = a' - a, \quad \nu_{\mathbb{B}} = b' - b, \quad \ldots$$

For example, for $2\mathbb{A} + \mathbb{B} \rightleftharpoons 2\mathbb{C}$ these are $\nu_{\mathbb{A}} = -2$, $\nu_{\mathbb{B}} = -1$, and $\nu_{\mathbb{C}} = +2$. Another example is the autocatalytic reaction $\mathbb{A} + \mathbb{B} \rightleftharpoons 3\mathbb{A}$; the stoichiometric coefficients for this transformation are $\nu_{\mathbb{A}} = +2$ and $\nu_{\mathbb{B}} = -1$. By mass conservation $\sum_{k} m_k \nu_k = 0$.[†]

Extent of Reaction

The Gibbs equation allows us to express changes of the internal energy, U, in terms of changes of the state variables. For example, for a closed, reactive system with three species ($\mathbb{A}$, $\mathbb{B}$, and $\mathbb{C}$) and the transformation (e.g., chemical reaction)

$$2\mathbb{A} + \mathbb{B} \rightleftharpoons 2\mathbb{C}$$

the Gibbs equation is

$$dU = TdS - pdV + \mu_{\mathbb{A}} d_{\mathrm{i}} M_{\mathbb{A}} + \mu_{\mathbb{B}} d_{\mathrm{i}} M_{\mathbb{B}} + \mu_{\mathbb{C}} d_{\mathrm{i}} M_{\mathbb{C}}.$$

[†]Neglecting relativistic effects.

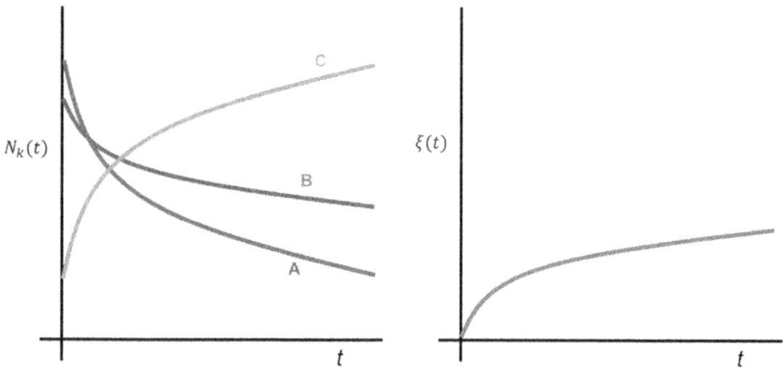

Figure 9.1: Graphs of $N_k(t)$ and $\xi(t)$ versus time for the reaction $2A+B \rightleftharpoons 2C$ for $d\xi/dt > 0$.

An alternative form uses mole number, N, instead of mass, M, to write

$$dU = TdS - pdV + \hat{\mu}_A d_i N_A + \hat{\mu}_B d_i N_B + \hat{\mu}_C d_i N_C$$

where $\hat{\mu}_k = m_k N_A \mu_k$.

But clearly these changes in mole number are not independent. In fact they are closely related, specifically for $2A + B \rightleftharpoons 2C$ we have

$$-\tfrac{1}{2}d_i N_A = -d_i N_B = \tfrac{1}{2}d_i N_C$$

That is, for the forward reaction the transformation consumes two moles of A and one mole of B for every two moles of C produced (and vice versa for the reverse reaction).

A simple way to handle this bookkeeping is to use a new state variable, ξ, introduced by De Donder called the *extent of reaction*. It is defined as

$$d\xi = \frac{1}{\nu_k}d_i N_k \qquad \text{so} \qquad d_i N_k = \nu_k d\xi$$

where ν_k are the stoichiometric coefficients for the reaction. Note that ξ is an extensive thermodynamic variable with units of moles. In our example, when $\xi = 3$ moles the transformation has consumed 6 moles of A and 3 moles of B to produce 6 moles of C.[‡]

In closed systems with only a single transformation we can use ξ as a state variable instead of $\mathbf{M}$ or $\mathbf{N}$. Using extent of reaction we write $U(S, V, \xi)$ with $\mathbf{N} = \mathbf{N_0} + \boldsymbol{\nu}\xi$ where the reference state $\mathbf{N_0}$ is when $\xi = 0$ (see Fig. 9.1). Using ξ simplifies the Gibbs equation to

$$dU = TdS - pdV - \mathcal{A}d\xi$$

[‡]If $\xi < 0$ then A and B are produced and C is consumed.

where

$$A = -\sum_{k=1}^{K} \nu_k \hat{\mu}_k = -\boldsymbol{\nu} \cdot \hat{\boldsymbol{\mu}}$$

is the *chemical affinity*. In our example, for the reaction $2A + B \rightleftharpoons 2C$

$$A = 2\hat{\mu}_A + \hat{\mu}_B - 2\hat{\mu}_C$$

The term "chemical" here is somewhat misleading in that we have the same form for any material transformation, such as changes of phase.

If there's more than one transformation occurring in a system then we have a separate ξ for each of them. For example, if we have four species and these two chemical reactions

$$
\begin{array}{rcll}
A + 2B & \rightleftharpoons & C & \text{(Reaction 1)} \\
A + C & \rightleftharpoons & D & \text{(Reaction 2)}
\end{array}
$$

then the Gibbs equation is

$$dU = TdS - pdV - (A_1 d\xi_1 + A_2 d\xi_2).$$

The chemical affinities are

$$A_1 = \hat{\mu}_A + 2\hat{\mu}_B - \hat{\mu}_C \qquad \text{and} \qquad A_2 = \hat{\mu}_A + \hat{\mu}_C - \hat{\mu}_D$$

with

$$\nu_{A,1} = -1, \qquad \nu_{B,1} = -2, \qquad \nu_{C,1} = +1, \qquad \nu_{D,1} = 0$$

$$\nu_{A,2} = -1, \qquad \nu_{B,2} = 0, \qquad \nu_{C,2} = -1, \qquad \nu_{D,2} = +1$$

being the stoichiometric coefficients for these reactions. The chemical affinity is a crucial element in the formulation of the Second Law for open and reactive systems, as we'll see in the next section.

9.3   Entropy in Homogeneous Systems

We now turn our attention to isolated systems that are homogeneous. The Gibbs equation in the entropy formulation is

$$dS = \frac{1}{T}dU + \frac{p}{T}dV - \frac{1}{T}\boldsymbol{\mu} \cdot d\mathbf{M}.$$

For an isolated system $dU = dV = d_e S = d_e M_k = 0$. If the system is inert, that is, if there are no transformations such as phase changes or chemical reactions, then we also have $d_i M_k = 0$ so $d\mathbf{M} = 0$. The Gibbs equation is then reduced to simply

$$d_i S = 0 \qquad \text{(Inert systems)}.$$

In other words, an inert, homogeneous, isolated system can *only* be at thermodynamic equilibrium. On the other hand, for reactive systems ($d_i M_k \neq 0$) the Gibbs equation is

$$d_i S = -\frac{1}{T}\boldsymbol{\mu} \cdot d_i \mathbf{M} = -\frac{1}{T}\sum_{k=1}^{K} \mu_k d_i M_k \qquad \text{(Reactive systems).}$$

In an isolated, homogeneous system the entropy can change by material transformations.

Let's work through an example of a homogeneous system that has three species, $\mathbb{A}$, $\mathbb{B}$ and $\mathbb{C}$ with a single transformation

$$2\mathbb{A} + \mathbb{B} \rightleftharpoons 2\mathbb{C}$$

The internal change of entropy due to this transformation is

$$d_i S = -\frac{1}{T}(\mu_\mathbb{A} d_i M_\mathbb{A} + \mu_\mathbb{B} d_i M_\mathbb{B} + \mu_\mathbb{C} d_i M_\mathbb{C})$$

$$= -\frac{1}{T}(\hat{\mu}_\mathbb{A} d_i N_\mathbb{A} + \hat{\mu}_\mathbb{B} d_i N_\mathbb{B} + \hat{\mu}_\mathbb{C} d_i N_\mathbb{C})$$

since $\hat{\mu}_k = m_k N_A \mu_k$. Using extent of reaction we write this as

$$d_i S = \frac{1}{T}(2\hat{\mu}_\mathbb{A} + \hat{\mu}_\mathbb{B} - 2\hat{\mu}_\mathbb{C})\, d\xi = \frac{\mathcal{A}}{T}\, d\xi$$

where $\mathcal{A}$ is the chemical affinity for this transformation. The rate of entropy production is

$$\frac{d_i S}{dt} = \left(\frac{\mathcal{A}}{T}\right)\left(\frac{d\xi}{dt}\right)$$

so we identify the thermodynamic force and thermodynamic flow to be,

$$\mathcal{F}_\xi = \frac{\mathcal{A}}{T} \qquad \text{and} \qquad \mathcal{J}_\xi = \frac{d\xi}{dt}.$$

Note that $\mathcal{J}_\xi$ is also called the *reaction velocity*.

When the system is out of equilibrium the transformation progresses in a direction determined by the affinity. If $\mathcal{A} > 0$ then $d\xi/dt > 0$ since by the Second Law $d_i S/dt = \mathcal{F}_\xi \mathcal{J}_\xi > 0$ so the transformation goes in the forward direction (consuming $\mathbb{A}$ and $\mathbb{B}$ to produce $\mathbb{C}$). As the transformation advances the composition (e.g., mass fractions M_k/M) changes, which changes the state variables (T, p, and μ_k) and thus the affinity also changes.

Eventually the entropy production goes to zero once the system reaches *chemical equilibrium*, which for a homogeneous system implies thermodynamic equilibrium.[§] For inhomogeneous systems to be at thermodynamic

---

[§]Chemical equilibrium is not required for the equations of state $p(U, V, \mathbf{M})$ and $T(U, V, \mathbf{M})$ to be well-defined.

equilibrium also requires thermal and mechanical equilibrium (see Section 7.3). In general, thermodynamic equilibrium requires that $d_i S/dt = 0$ for *all* thermodynamic forces and flows.

Example 9.2: A homogeneous system has the transformation, $2A + B \rightleftharpoons 2C$. Initially the molar chemical potentials are $\hat{\mu}_A = 3$ kJ/mol, $\hat{\mu}_B = 1$ kJ/mol, and $\hat{\mu}_C = 4$ kJ/mol. Is the system at thermodynamic equilibrium? If not, is species C produced or consumed?

Solution: The chemical affinity is

$$A = 2\hat{\mu}_A + \hat{\mu}_B - 2\hat{\mu}_C = -1 \text{ kJ/mol}.$$

Since $A \neq 0$ the system is not at equilibrium. Given that $d_i S/dt = \mathcal{F}_\xi \mathcal{J}_\xi > 0$ since $\mathcal{F}_\xi = A/T < 0$ then $\mathcal{J}_\xi < 0$. The flow being negative means $d_i N_C/dt = 2d\xi/dt < 0$ so C is consumed, producing A and B. Note that as the reaction proceeds the values of the chemical potentials will change since they depend on temperature, pressure, and composition with $A \to 0$ as the system approaches equilibrium.

Finally, when there's more than one transformation (e.g., multiple chemical reactions) each has its own chemical affinity, A_r, and its own extent of reaction ξ_r. We write,

$$\frac{d_i S}{dt} = \sum_r \mathcal{F}_r \mathcal{J}_r = \sum_r \left(\frac{A_r}{T}\right)\left(\frac{d\xi_r}{dt}\right)$$

where the sum is over all reactions and $A_r = -\boldsymbol{\nu}_r \cdot \hat{\boldsymbol{\mu}}$ is the chemical affinity for reaction r.

9.4   Heat for Open Systems*

In Section 5.1 the relation between entropy and heat was presented as

$$dQ = T d_e S \qquad \text{(Closed systems)}.$$

This relation applies to closed systems and we now generalize it to open systems for which $\mathbf{M}$ can vary by mass entering or leaving the system through a porous boundary.

The general form of the Gibbs equation in the energy representation is

$$dU(S, V, \mathbf{M}) = T dS - p dV + \boldsymbol{\mu} \cdot d\mathbf{M}.$$

Also recall that we express the change in mass as, $d\mathbf{M} = d_e \mathbf{M} + d_i \mathbf{M}$ where $d_e \mathbf{M}$ is the change due to external sources (e.g., mass diffusing into the system) and $d_i \mathbf{M}$ is from internal sources (e.g., chemical reactions). The entropy change is also written as $dS = d_e S + d_i S$ so

$$dU = [T d_e S + \boldsymbol{\mu} \cdot d_e \mathbf{M}] - p dV + [T d_i S + \boldsymbol{\mu} \cdot d_i \mathbf{M}]$$

by collecting the external and internal contributions.

In Section 9.3 we showed that a homogeneous system can produce entropy by internal processes if $\mathbf{M}$ changes

$$d_i S = -\frac{1}{T}\boldsymbol{\mu} \cdot d_i \mathbf{M}.$$

Putting this into the previous equation gives

$$dU = [T d_e S + \boldsymbol{\mu} \cdot d_e \mathbf{M}] - p dV.$$

Finally, comparing with the First Law, $dU = dQ + dW = dQ - pdV$ gives

$$dQ = T d_e S + \boldsymbol{\mu} \cdot d_e \mathbf{M} \qquad \text{(Open systems)}.$$

Notice the additional term in dQ due to the flow of matter allowed in open systems. For closed systems $d_e \mathbf{M} = 0$ and we recover our previous result for heat, $dQ = T d_e S$. And I should mention that some authors (e.g., Callen) refer to the $\boldsymbol{\mu} \cdot d_e \mathbf{M}$ term as "chemical work" and put it into the definition of dW instead of into dQ. Potato, Potahto.

$$* * *$$

This chapter introduced chemical potential as an equation of state and chemical affinity as a thermodynamic force for transformations. In the next chapter we'll apply the Second Law to analyze more complicated scenarios involving reactive and open systems.

Chapter 10

Second Law – Open & Reactive Systems*

Having extended Gibbs equation to reactive and open systems we now apply the Second Law to some important examples. The first case is a reactive system that models a fundamental metabolic mechanism in living systems. In the second we study the diffusion of matter in an open system. In both examples we'll see how these the irreversible evolution of processes can be manipulated without violating the Second Law.

10.1 Endergonic Reactions

Let's apply the results in Section 9.3 to model a mechanism that's essential to the operation of living systems. Consider a homogeneous isolated system with the reaction

$$X \rightleftharpoons Y \qquad \text{(Reaction 1)}$$

For this reaction the entropy production rate is

$$\frac{d_i S}{dt} = \frac{\mathcal{A}_1}{T}\frac{d\xi_1}{dt} = \frac{\mathcal{A}_1}{T}\frac{d_i N_Y}{dt} \geq 0.$$

Let's say that $\hat{\mu}_X > \hat{\mu}_Y$ so the chemical affinity $\mathcal{A}_1 = \hat{\mu}_X - \hat{\mu}_Y > 0$. This means that $d_i N_Y/dt > 0$, that is, Y is produced while X is consumed. But suppose we want to produce X instead of consume it? How can we do this without violating the Second Law?

The key is to introduce a second reaction

$$F \rightleftharpoons W \qquad \text{(Reaction 2)}$$

for which $\mathcal{A}_2 = \hat{\mu}_F - \hat{\mu}_W > 0$ so F is consumed and W is produced (i.e., $\hat{\mu}_F > \hat{\mu}_W$). If these two elementary reactions occur independently then this extra reaction doesn't help. However, if they're coupled then we can achieve our goal of producing X.

We introduce two new species, Z and Z^*, with Z^* being the "high-energy version" of Z, specifically, $\hat{\mu}_{Z^*} > \hat{\mu}_Z$. The reactions are modified to be

$$Y + Z^* \;\rightleftharpoons\; X + Z \qquad\qquad \text{(Reaction } 1')$$
$$F + Z \;\rightleftharpoons\; W + Z^* \qquad\qquad \text{(Reaction } 2').$$

Notice that when Reaction $1'$ goes in the forward direction it produces X, as desired. In chemistry this is called an *endergonic reaction*, from the Greek prefix endo- ("within") and the Greek word ergon ("work"), with species Z^* doing the work, so to speak, to drive the reaction.

The total entropy production rate for these reactions is

$$\frac{d_i S}{dt} = \frac{\mathcal{A}_{1'}}{T}\frac{d\xi_{1'}}{dt} + \frac{\mathcal{A}_{2'}}{T}\frac{d\xi_{2'}}{dt}$$

where

$$\mathcal{A}_{1'} = \hat{\mu}_Y + \hat{\mu}_{Z^*} - \hat{\mu}_X - \hat{\mu}_Z \qquad \text{and} \qquad \mathcal{A}_{2'} = \hat{\mu}_F + \hat{\mu}_Z - \hat{\mu}_W - \hat{\mu}_{Z^*}.$$

Defining $\Delta\hat{\mu}_Z = \hat{\mu}_{Z^*} - \hat{\mu}_Z$ gives

$$\mathcal{A}_{1'} = \Delta\hat{\mu}_Z - \mathcal{A}_1 \qquad \text{and} \qquad \mathcal{A}_{2'} = \mathcal{A}_2 - \Delta\hat{\mu}_Z.$$

With the appropriate values for the chemical potentials both reactions can go in the forward direction, that is, we can have $\mathcal{A}_{1'} > 0$ and $\mathcal{A}_{2'} > 0$.

Example 10.1: Given the chemical potentials (in kJ/mol)

$$\hat{\mu}_X = 8; \qquad \hat{\mu}_Y = 6; \qquad \hat{\mu}_F = 5; \qquad \hat{\mu}_W = 1; \qquad \hat{\mu}_Z = 3; \qquad \hat{\mu}_{Z^*} = 6$$

find the affinities of the above reactions (Reactions $1, 2, 1'$, and $2'$).
Solution: By arithmetic, $\Delta\hat{\mu}_Z = 3$ so $\mathcal{A}_1 = 2$, $\mathcal{A}_2 = 4$, $\mathcal{A}_{1'} = 1$, $\mathcal{A}_{2'} = 1$. All these affinities are positive so all the reactions go in the forward direction. The total chemical potentials for reactants and products are illustrated as an "energy level" diagram in Figure 10.1.

If the consumption of species Z^* by the first reaction is balanced by the production of it by the second reaction then species Z^* and Z are at steady state ($dN_{Z^*}/dt = dN_Z/dt = 0$). In that case the net reaction is

$$Y + F \rightleftharpoons X + W \qquad\qquad \text{(Net Reaction)}$$

with chemical affinity

$$\mathcal{A}^{\text{net}} = \mu_Y + \mu_F - \mu_X - \mu_W = \mathcal{A}_{1'} + \mathcal{A}_{2'} = \mathcal{A}_2 - \mathcal{A}_1.$$

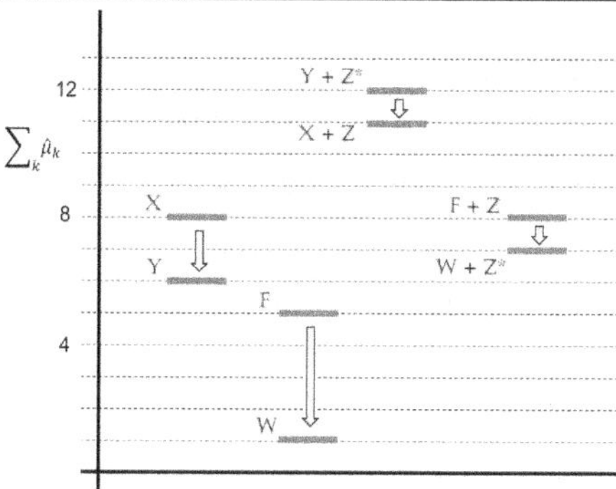

Figure 10.1: Total chemical potentials for reactants and products for Reactions $1, 2, 1'$, and $2'$. Note that the stoichiometric coefficients are ± 1 in this example.

When $\mathbb{Z}$ and $\mathbb{Z}^*$ are at steady state $d\xi_{1'}/dt = d\xi_{2'}/dt$ so the entropy production for the coupled reactions is

$$\frac{d_i S}{dt} = \frac{\mathcal{A}_{1'} + \mathcal{A}_{2'}}{T}\frac{d\xi}{dt} = \frac{\mathcal{A}^{\text{net}}}{T}\frac{d\xi}{dt}.$$

With the appropriate values of chemical potentials we have $\mathcal{A}^{\text{net}} > 0$ achieving the desired goal of producing $\mathbb{X}$ and consuming $\mathbb{Y}$.

Biological systems use this trick with ADP (Adenosine diphosphate) and ATP (Adenosine triphosphate) playing the roles of $\mathbb{Z}$ and $\mathbb{Z}^*$. The energy for converting $\mathbb{Z}$ into $\mathbb{Z}^*$ is obtained by turning food (species $\mathbb{F}$) into waste (species $\mathbb{W}$), by a process known as *catabolism*.Just as a refrigerator uses work to transfer heat energy against its "thermal affinity" (from cold to hot), in this example endergonic reactions drive the conversion of $\mathbb{Y}$ into $\mathbb{X}$, going against the chemical affinity of the uncoupled reaction. In biological systems this is called *anabolism*(from the Greek "upward" and "to throw"). The details of these reaction processes are complicated but the essential thermodynamics is the same as in these simple examples.

10.2   Diffusion

Consider a pure, heterogenous, isolated system similar to the one discussed in Section 7.3 but now the two inert subsystems can exchange both energy *and* mass. An example would be an isolated container of helium gas divided into left and right sides by a stationary, porous partition (see Figure 10.2).

The subsystem volumes, V_L and V_R, are fixed so the Gibbs equation for

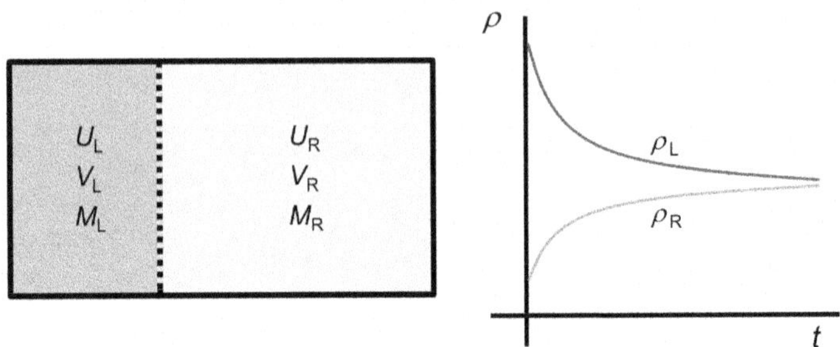

Figure 10.2:   Mass diffusion in a heterogeneous isolated system: (left) Schematic illustration of the system; (right) Mass density $\rho_{\mathrm{L}}(t)$ and $\rho_{\mathrm{R}}(t)$ versus time for an ideal gas.

the Left subsystem is

$$dS_{\mathrm{L}} = \frac{1}{T_{\mathrm{L}}} dU_{\mathrm{L}} - \frac{1}{T_{\mathrm{L}}} \mu_{\mathrm{L}} dM_{\mathrm{L}}$$

and similarly for the Right subsystem. For this isolated system $S = S_{\mathrm{L}} + S_{\mathrm{R}}$ and $dS = d_{\mathrm{i}}S$ since $d_{\mathrm{e}}S = 0$. The rate of entropy production is

$$\frac{d_{\mathrm{i}}S}{dt} = \left[\frac{1}{T_{\mathrm{L}}} \frac{dU_{\mathrm{L}}}{dt} - \frac{1}{T_{\mathrm{L}}} \mu_{\mathrm{L}} \frac{dM_{\mathrm{L}}}{dt} \right] + \left[\frac{1}{T_{\mathrm{R}}} \frac{dU_{\mathrm{R}}}{dt} - \frac{1}{T_{\mathrm{R}}} \mu_{\mathrm{R}} \frac{dM_{\mathrm{R}}}{dt} \right]$$

or

$$\frac{d_{\mathrm{i}}S}{dt} = \left(\frac{1}{T_{\mathrm{R}}} - \frac{1}{T_{\mathrm{L}}} \right) \left(\frac{dU_{\mathrm{R}}}{dt} \right) + \left(\frac{\mu_{\mathrm{L}}}{T_{\mathrm{L}}} - \frac{\mu_{\mathrm{R}}}{T_{\mathrm{R}}} \right) \left(\frac{dM_{\mathrm{R}}}{dt} \right)$$

since by conservation of mass and energy $dM_{\mathrm{L}} = -dM_{\mathrm{R}}$ and $dU_{\mathrm{L}} = -dU_{\mathrm{R}}$.

In this system there are two thermodynamic flows (energy and mass) and two corresponding thermodynamic forces. We write this as

$$\frac{d_{\mathrm{i}}S}{dt} = \mathcal{F}_U \, \mathcal{J}_U + \mathcal{F}_M \, \mathcal{J}_M$$

where the two thermodynamic forces for energy and mass are

$$\mathcal{F}_U = \frac{1}{T_{\mathrm{R}}} - \frac{1}{T_{\mathrm{L}}} \qquad\qquad \mathcal{F}_M = \frac{\mu_{\mathrm{L}}}{T_{\mathrm{L}}} - \frac{\mu_{\mathrm{R}}}{T_{\mathrm{R}}}$$

with the corresponding thermodynamic flows

$$\mathcal{J}_U = \frac{dU_{\rightarrow}}{dt} \qquad\qquad \mathcal{J}_M = \frac{dM_{\rightarrow}}{dt}.$$

From this result we see that at thermodynamic equilibrium $T_{\mathrm{R}} = T_{\mathrm{L}}$ and $\mu_{\mathrm{R}} = \mu_{\mathrm{L}}$.

Example 10.2: Show that for a pure ideal gas the heterogenous system in Fig. 10.2 has equal density in the two subsystems (i.e., $M_R/V_R = M_L/V_L$) at equilibrium. **Solution:** Using the result from Sections 5.2 and 9.1, the chemical potential for an ideal gas is

$$\mu(T, V, M) = -\frac{1}{M} TS + c_p T = \frac{k_B T}{m} \ln(M/V) + f(T)$$

where $f(T)$ is a function only of temperature. At equilibrium $T_R = T_L$ and $\mu(T_R, V_R, M_R) = \mu(T_L, V_L, M_L)$ so, by inspection, the density $\rho = M/V$ is the same on both sides.

In the example above the equilibrium state, $\rho_L = \rho_R$, is simple since the chemical potential for an ideal gas is simple. The situation can be more complicated for mixtures, such as oil and water, where $\mu_k(T, V, M_{oil}, M_{water})$. In such cases the equilibrium conditions

$$\mu_{oil,L} = \mu_{oil,R} \qquad \text{and} \qquad \mu_{water,L} = \mu_{water,R}$$

can result in $\rho_{oil,L} \neq \rho_{oil,R}$. In other words, the oil and water mixture can separate because that state has higher entropy than a uniform, homogeneous state. The situation is similar for phase separation, such as in mixtures of liquid water and solid ice.

Diffusion as a Transformation

Instead of describing diffusion in terms of M one can use N, the number of moles. For a pure system, we then write the thermodynamic force and flow of matter as

$$\mathcal{F}_N = \frac{\hat{\mu}_L}{T_L} - \frac{\hat{\mu}_R}{T_R} \qquad \text{and} \qquad \mathcal{J}_N = \frac{dN_\rightarrow}{dt}$$

with $d_i S/dt = \mathcal{F}_U \mathcal{J}_U + \mathcal{F}_N \mathcal{J}_N$. If the temperatures of the subsystems are equal ($T_L = T_R = T$) then the thermodynamic force and flow of matter may be expressed as

$$\mathcal{F}_\xi = \frac{\mathcal{A}}{T} \qquad \text{where} \qquad \mathcal{A} = \hat{\mu}_L - \hat{\mu}_R$$

and

$$\mathcal{J}_\xi = \frac{d\xi}{dt} \qquad \text{where} \qquad \xi(t) = N_R(t) - N_R(0).$$

Here the extent of reaction, ξ is defined such that $\xi = 0$ at time $t = 0$. In essence this is equivalent to treating diffusion as the transformation, $\mathbb{A}_L \rightleftharpoons \mathbb{A}_R$, that is, when a molecule diffuses from the Left subsystem to the Right it transforms from type $\mathbb{A}_L$ to type $\mathbb{A}_R$. This result illustrates that in thermodynamics there is a close connection between material transformations in homogeneous systems and the diffusion of matter in heterogeneous systems.

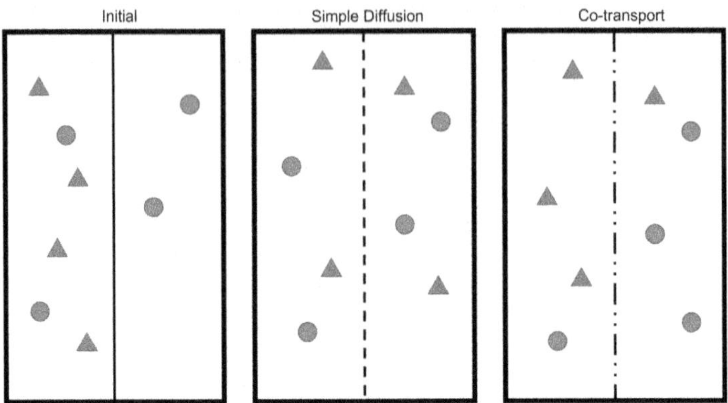

Figure 10.3: An example illustrating simple diffusion versus co-transport; triangles and circles represent species $\mathbb{A}$ and $\mathbb{B}$, respectively

Formulating diffusion as a material transformation is a useful way to model transport in biological cells.* For example, some transporter proteins use *co-transport* (also known as symport) to allow certain species to enter and exit the cell but only when they do so together. For two species, $\mathbb{A}$ and $\mathbb{B}$, that co-transport between Left and Right subsystems (inside and outside the cell) the material transformation is

$$\mathbb{A}_L + \mathbb{B}_L \rightleftharpoons \mathbb{A}_R + \mathbb{B}_R.$$

The chemical affinity is

$$\mathcal{A} = \hat{\mu}_{\mathbb{A}_L} + \hat{\mu}_{\mathbb{B}_L} - \hat{\mu}_{\mathbb{A}_R} - \hat{\mu}_{\mathbb{B}_R}.$$

The chemical potential in a dilute solution is similar to that of a dilute (ideal) gas, that is,

$$\hat{\mu}_k(T, V, N_k) = \hat{\mu}_k^{\circ}(T) + RT \ln(N_k/V).$$

With this form for $\hat{\mu}$ at thermodynamic equilibrium ($\mathcal{A} = 0$) the state of the system with co-transport diffusion is

$$[\mathbb{A}_L][\mathbb{B}_L] = [\mathbb{A}_R][\mathbb{B}_R]$$

where $[k] = N_k/V$ is the molar concentration, typically given as molarity M (moles per liter).

Example 10.3:   Figure 10.3 illustrates a heterogeneous system (with $V_L = V_R$) containing a dilute solution. The initial molar concentrations are:

$$[\mathbb{A}_L]_0 = 4 \text{ M}; \qquad [\mathbb{A}_R]_0 = 0; \qquad [\mathbb{B}_L]_0 = [\mathbb{B}_R]_0 = 2 \text{ M}.$$

---

* "Key biology you should have learned in physics class: Using ideal-gas mixtures to understand biomolecular machines", Daniel M. Zuckerman, *Am. J. Phys.* **88**, 182 (2020).

Find the equilibrium state when mass transfer occurs by simple diffusion and when it occurs by co-transport.

Solution: By mass conservation $\Delta N_{A_L} = -\Delta N_{A_R}$ so $\Delta[A_L] = -\Delta[A_R]$ since the subsystems have equal volumes. For simple diffusion the equilibrium state is $[A_L] = [A_R]$ and $[B_L] = [B_R]$ so $\Delta[A_L] = -2$ M and $\Delta[B_L] = 0$. With this, the equilibrium state is

$$[A_L] = [A_R] = [B_L] = [B_R] = 2 \text{ M} \qquad \text{(Simple Diffusion)}.$$

With co-transport $\Delta[A_L] = \Delta[B_L] = \Delta_L$ so the equilibrium state, $[A_L][B_L] = [A_R][B_R]$, is

$$([A_L]_0 + \Delta_L)([B_L]_0 + \Delta_L) = ([A_R]_0 - \Delta_L)([B_R]_0 - \Delta_L).$$

Solving this we get $\Delta_L = -1$ M so

$$
\begin{aligned}
[A_L] &= 3 \text{ M} & [A_R] &= 1 \text{ M} \\
[B_L] &= 1 \text{ M} & [B_R] &= 3 \text{ M} & \text{(Co} - \text{transport)}.
\end{aligned}
$$

With co-transport the initial excess of A in the Left subsystem (e.g., outside the cell) causes B molecules to be transported into the Right subsystem (into the cell) creating a concentration of B that's three times larger on the Right than on the Left. The entropy for species B is lowered without violating the Second Law by having an increase in the entropy for species A.

$$* \; * \; *$$

For many applications we've found it useful to express the rate of entropy production as $d_i S/dt = \mathcal{J}\mathcal{F} \geq 0$ where $\mathcal{J}$ is a thermodynamic flow and $\mathcal{F}$ is a thermodynamic force. In the next chapter we'll see that when a system is not far from equilibrium we may write a linear relation between $\mathcal{J}$ and $\mathcal{F}$. Ohm's law is an example you're already familiar with and we'll see that there are many other such phenomenological laws.

Chapter 11

Phenomenological Laws

In Chapter 7 we formulated entropy production in terms of thermodynamic forces and flows. In this chapter we link these forces and flows by introducing phenomenological laws. The special properties of the coefficients appearing in these laws, including Onsager reciprocity, is highlighted.

11.1 Phenomenological Laws

We start with the example from Section 7.3, specifically, an isolated heterogeneous system consisting of two subsystems (Left and Right) at temperatures T_L and T_R. As we saw the rate of entropy production is $d_i S/dt = \mathcal{F}_U \, \mathcal{J}_U$ with

$$\mathcal{F}_U \;=\; \frac{1}{T_R} - \frac{1}{T_L} \qquad \text{(Thermodynamic Force)}$$

$$\mathcal{J}_U \;=\; \frac{dU_\rightarrow}{dt} \qquad \text{(Thermodynamic Flow)}.$$

We've treated the force and the flow as independent, yet from experience we know that the greater the temperature difference, the faster energy flows from hot to cold. In fact, *Newton's Law of Cooling* states that the rate at which energy goes from Left to Right is

$$\frac{dU_\rightarrow}{dt} = -\mathcal{C}\,(T_R - T_L)$$

where the coefficient $\mathcal{C}$ characterizes the heat conductivity; since heat energy goes from hot to cold $\mathcal{C} \geq 0$. When the difference in temperatures is small the thermodynamic force is

$$\mathcal{F}_U = \frac{1}{T_R} - \frac{1}{T_L} \approx -\frac{1}{T^2}(T_R - T_L)$$

where $T = \frac{1}{2}(T_L + T_R)$ is the average temperature. By comparison we see that

$$\mathcal{J}_U = \frac{dU_\rightarrow}{dt} = -\mathfrak{C}(T_R - T_L) \approx -\mathfrak{C}(-T^2 \mathcal{F}_U)$$

so

$$\mathcal{J}_U \approx L\mathcal{F}_U \qquad \text{where} \qquad L = \mathfrak{C}T^2$$

is called the *phenomenological coefficient* or *Onsager coefficient*. In words, the flow is proportional to the force with the constant of proportionality being the coefficient L. This linear relation between $\mathcal{J}$ and $\mathcal{F}$ works well for many processes, such as heat conduction and mass diffusion, though for chemical reactions an exponential relation is more accurate.[*]

Another way of arriving at this relation is to formulate the flow as a function of the force and doing the Taylor expansion

$$\mathcal{J}(\mathcal{F}) = \mathcal{J}(0) + \frac{d\mathcal{J}}{d\mathcal{F}}\bigg|_{\mathcal{F}=0} \mathcal{F} + \ldots \approx \frac{d\mathcal{J}}{d\mathcal{F}}\bigg|_{\mathcal{F}=0} \mathcal{F}$$

since $\mathcal{J} = 0$ at $\mathcal{F} = 0$. Having made the linear approximation of dropping the quadratic and higher order terms we have

$$\mathcal{J} = L\mathcal{F} \qquad \text{where} \qquad L = \frac{d\mathcal{J}}{d\mathcal{F}}\bigg|_{\mathcal{F}=0}.$$

This is called the linear *phenomenological law* relating the flow to the force.

Sometimes it's useful to write the phenomenological law as,

$$\mathcal{F} = P\mathcal{J} \qquad \text{where} \qquad P = \frac{1}{L} = \frac{d\mathcal{F}}{d\mathcal{J}}\bigg|_{\mathcal{J}=0}.$$

If we interpret the Onsager coefficient L as a conductance then P is interpreted as a resistance. In fact, Ohm's law happens to be an example of a linear phenomenological law, as shown in the next example.

Writing $\mathcal{J} = L\mathcal{F}$, the rate of entropy production is

$$\frac{d_i S}{dt} = \mathcal{J}\mathcal{F} = (L\mathcal{F})\mathcal{F} = L\mathcal{F}^2$$

and this applies in the linear regime, that is, when the flow is linearly proportional to the force. Similarly we can write

$$\frac{d_i S}{dt} = \mathcal{J}\mathcal{F} = \mathcal{J}(P\mathcal{J}) = P\mathcal{J}^2.$$

Since the Second Law requires that $d_i S/dt \geq 0$ then $L \geq 0$, $P \geq 0$.

---

[*]D. Venerus and H. Öttinger, *A Modern Course in Transport Phenomena*, Cambridge Univ. Press (2018).

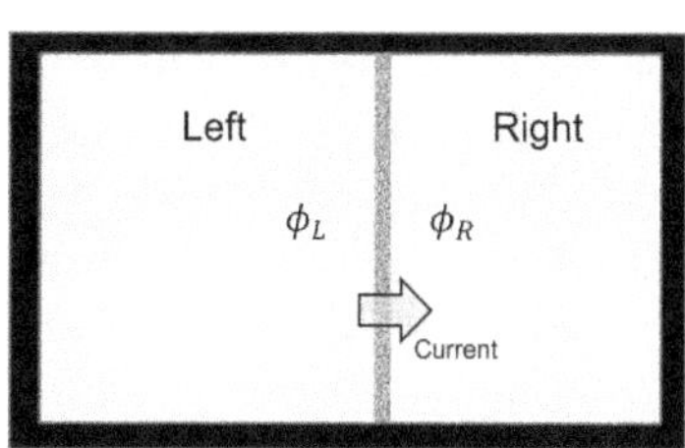

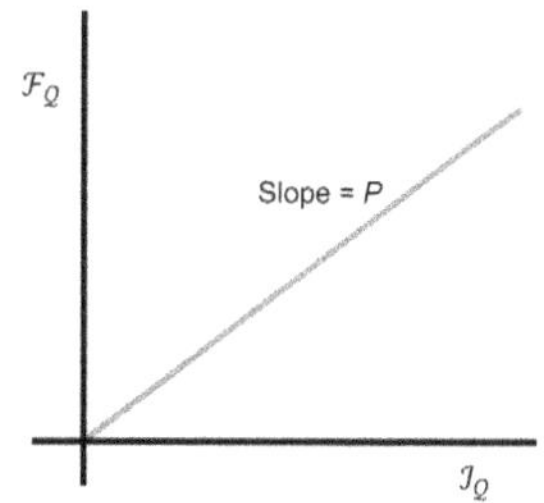

Figure 11.1: Heterogeneous system with subsystems at different electric potential with an electric current passing between them (e.g., electrolyte solutions separated by a porous membrane). The phenomenological law, $\mathcal{F}_Q = P\mathcal{J}_Q$, is Ohm's law.

Example 11.1: For an electrical system we may write the Gibbs equation as $dU = TdS + \phi dQ$ where ϕ is the electric potential and Q is the electric charge.[†] For a heterogeneous system at constant T with potentials ϕ_L and ϕ_R (see Fig. 11.1) find the rate of entropy production in terms of the electric current.
Solution: The thermodynamic flow of charge (i.e., electric current) is $\mathcal{J}_Q = dQ_{\rightarrow}/dt$. This current is driven by a potential difference, that is, the thermodynamic force

$$\mathcal{F}_Q = \frac{\phi_L}{T_L} - \frac{\phi_R}{T_R} = -\frac{\phi_R - \phi_L}{T}$$

for constant temperature. The phenomenological law is *Ohm's law*, which in modern thermodynamics is written as

$$\mathcal{F}_Q = P\mathcal{J}_Q$$

where $P = \mathfrak{R}/T$ is the Onsager coefficient and $\mathfrak{R}$ is the ohmic resistance. The rate of entropy production is

$$\frac{d_i S}{dt} = \mathcal{F}_Q\mathcal{J}_Q = \frac{\mathfrak{R}}{T}\mathcal{J}_Q^2.$$

We see that $\mathfrak{R} \geq 0$ is required by the Second Law. Note that $T d_i S/dt = \mathfrak{R}\mathcal{J}_Q^2$, which is Joule's law for the electric power for ohmic heating from an electric current.

Two final points: First, phenomenological coefficients cannot be derived from thermodynamic state variables. In other words, while it's possible to use $S(U, V, M)$ to obtain equations of state (T, p, μ) and thermodynamic coefficients $(c_V, \alpha, \kappa_T, \text{etc.})$ it will *not* give any information regarding transport properties such as thermal conductivity or electrical resistance. And equilibrium statistical mechanics is no help either. To determine phenomenological coefficients have to resort to other types of modeling, such as kinetic theory, or measure them in computer simulations and laboratory experiments.

Second, it should be emphasized that phenomenological coefficients are usually *not* constants. In the example above the Onsager coefficient $P = \mathfrak{R}/T$

[†]Notice that analogy between ϕdQ and μdM.

and for many materials the electrical resistance also varies with T so $P(T)$ may be a complicated function of temperature. In general it could also be a function of pressure and, for a system with multiple species, concentration (e.g., electrical resistance of seawater varies with salt concentration). What we can say is that $P(T, p, \mathbf{M}) \geq 0$ and it does not depend on gradients of the thermodynamic variables, with similar results for the Onsager coefficients in continuous systems.

11.2   Cross-coupled Forces and Flows*

Onsager theory becomes much more interesting and useful when there's more than one thermodynamic flow. In general, when there are $N_\Rightarrow$ thermodynamic flows, there are also $N_\Rightarrow$ thermodynamic forces. The phenomenological equations take the form

$$\mathcal{J}_i = L_{i,1}\mathcal{F}_1 + L_{i,2}\mathcal{F}_2 + \ldots + L_{i,N_\Rightarrow}\mathcal{F}_{N_\Rightarrow} = \sum_{j=1}^{N_\Rightarrow} L_{i,j}\mathcal{F}_j.$$

The rate of entropy production is

$$\frac{d_{\mathrm{i}}S}{dt} = \sum_{i=1}^{N_\Rightarrow} \mathcal{J}_i\mathcal{F}_i = \sum_{i=1}^{N_\Rightarrow}\sum_{j=1}^{N_\Rightarrow} L_{i,j}\,\mathcal{F}_i\mathcal{F}_j.$$

Defining the Onsager matrix, $\mathbf{L} = \{L_{i,j}\}$, we can write this as

$$\frac{d_{\mathrm{i}}S}{dt} = \boldsymbol{\mathcal{J}} \cdot \boldsymbol{\mathcal{F}} = (\mathbf{L}\,\boldsymbol{\mathcal{F}}) \cdot \boldsymbol{\mathcal{F}} = \boldsymbol{\mathcal{F}}^{\mathrm{T}}\mathbf{L}\,\boldsymbol{\mathcal{F}}$$

where $\boldsymbol{\mathcal{J}}$ and $\boldsymbol{\mathcal{F}}$ are column vectors and $\boldsymbol{\mathcal{F}}^{\mathrm{T}}$ is the transpose (i.e., $\boldsymbol{\mathcal{F}}$ as a row vector).

By the Second Law $d_{\mathrm{i}}S/dt \geq 0$ which means that the Onsager matrix is *positive semi-definite*. The definition of a positive semi-definite matrix is that for all non-zero column vectors $\mathbf{a}$ the product $\mathbf{a}^{\mathrm{T}}\mathbf{L}\mathbf{a} \geq 0$. Note that a matrix $\mathbf{L}$ is positive semi-definite if and only if it has all non-negative eigenvalues.

Example 11.2:   Show that the matrix $L_{1,1} = 1$, $L_{2,2} = 2$, $L_{1,2} = L_{2,1} = -1$ is positive semi-definite.

Solution: By the definition of positive semi-definite we require that

$$\mathbf{a}^{\mathrm{T}}\mathbf{L}\mathbf{a} = (a_1, a_2) \begin{pmatrix} 1 & -1 \\ -1 & 2 \end{pmatrix} \begin{pmatrix} a_1 \\ a_2 \end{pmatrix} \geq 0$$

for all values of a_1 and a_2. Evaluating the vector-matrix multiplication we have

$$(a_1, a_2) \begin{pmatrix} a_1 - a_2 \\ -a_1 + 2a_2 \end{pmatrix} = a_1^2 - a_1 a_2 - a_1 a_2 + 2a_2^2 = (a_1 - a_2)^2 + a_2^2.$$

By inspection this is non-negative for all values of a_1 and a_2. By the way, the eigenvalues of $\mathbf{L}$ are $(3 \pm \sqrt{5})/2$ so they are both positive. In general the matrix is positive semi-definite if $L_{1,1} \geq 0$, $L_{2,2} \geq 0$, and $4L_{1,1}L_{2,2} \geq (L_{1,2} + L_{2,1})^2$.

Seebeck and Peltier Effects

Let's look at an example with just two forces and two flows, specifically, consider a heterogeneous system where $\mathcal{J}_U$ is the flow of heat energy and $\mathcal{J}_Q$ is the flow of electric charge

$$\mathcal{J}_U = \frac{dU_\rightarrow}{dt} \quad \text{and} \quad \mathcal{J}_Q = \frac{dQ_\rightarrow}{dt}$$

where Q is the electric charge. The corresponding thermodynamic forces are

$$\mathcal{F}_U = \frac{1}{T_\mathrm{R}} - \frac{1}{T_\mathrm{L}} \quad \text{and} \quad \mathcal{F}_Q = \frac{\phi_\mathrm{L}}{T_\mathrm{L}} - \frac{\phi_\mathrm{R}}{T_\mathrm{R}}$$

where ϕ is the electric potential.

For this system the phenomenological laws are

$$\mathcal{J}_U = L_{UU}\mathcal{F}_U + L_{UQ}\mathcal{F}_Q$$
$$\mathcal{J}_Q = L_{QU}\mathcal{F}_U + L_{QQ}\mathcal{F}_Q$$

where L_{UU}, L_{UQ}, L_{QU}, and L_{QQ} are the phenomenological coefficients. When there's a difference of temperature with $\mathcal{F}_Q = 0$ then the flow of energy is $\mathcal{J}_U = L_{UU}\mathcal{F}_U$ and the coefficient L_{UU} is given by the thermal conductivity. Similarly, when there's a difference of potential (e.g., an applied voltage) but constant temperature then the electric current is $\mathcal{J}_Q = L_{QQ}\mathcal{F}_Q$ and the coefficient L_{QQ} is given by the electrical conductivity.

The interesting point is that, depending on the value of L_{QU}, we can have an electric current even when $\mathcal{F}_Q = 0$. This is known as the *Seebeck effect* and it is how a thermoelectric generator operates. Similarly, when there's a potential difference but no temperature difference then there is a heat flow, $\mathcal{J}_U = L_{UQ}\mathcal{F}_Q$, which is known as the *Peltier effect*; this is how a thermoelectric cooler operates.

Example 11.3: The pure resistance is defined as

$$P_Q^* = \frac{\mathcal{F}_Q}{\mathcal{J}_Q} \quad \text{when} \quad \mathcal{J}_U = 0.$$

Find an expression for P_Q^* in terms of L_{UU}, L_{UQ}, L_{QU}, and L_{QQ}.
Solution: From the phenomenological laws, when $\mathcal{J}_U = 0$

$$\mathcal{F}_U = -\frac{L_{UQ}}{L_{UU}}\mathcal{F}_Q.$$

Inserting this into the other phenomenological equation

$$\mathcal{J}_Q = L_{QU}\left(-\frac{L_{UQ}}{L_{UU}}\mathcal{F}_Q\right) + L_{QQ}\mathcal{F}_Q = \left\{L_{QQ} - \frac{L_{QU}L_{UQ}}{L_{UU}}\right\}\mathcal{F}_Q$$

so

$$P_Q^* = \frac{\mathcal{F}_Q}{\mathcal{J}_Q} = \frac{L_{UU}}{L_{QQ}L_{UU} - L_{UQ}L_{QU}}.$$

When the cross-coefficients are small $P_Q^* \approx 1/L_{QQ}$.

Soret and Dufour Effects

As a second example take a heterogeneous system in which energy and matter could be exchanged between the left and right sides. There's a thermodynamic force for the energy flow and a force for each species; for just two species, $\mathbb{X}$ and $\mathbb{Y}$, the forces are

$$\mathcal{F}_U = \frac{1}{T_R} - \frac{1}{T_L} \qquad \mathcal{F}_{\mathbb{X}} = \frac{\mu_{\mathbb{X},L}}{T_L} - \frac{\mu_{\mathbb{X},R}}{T_R} \qquad \mathcal{F}_{\mathbb{Y}} = \frac{\mu_{\mathbb{Y},L}}{T_L} - \frac{\mu_{\mathbb{Y},R}}{T_R}.$$

Similarly, the thermodynamic flows for energy and mass are

$$\mathcal{J}_U = \frac{dU_{\rightarrow}}{dt} \qquad \mathcal{J}_{\mathbb{X}} = \frac{dN_{\mathbb{X}\rightarrow}}{dt} \qquad \mathcal{J}_{\mathbb{Y}} = \frac{dN_{\mathbb{Y}\rightarrow}}{dt}.$$

In general, all thermodynamic forces may contribute to a thermodynamic flow so the phenomenological laws are

$$\begin{aligned}
\mathcal{J}_U &= L_{UU}\mathcal{F}_U + L_{U\mathbb{X}}\mathcal{F}_{\mathbb{X}} + L_{U\mathbb{Y}}\mathcal{F}_{\mathbb{Y}} \\
\mathcal{J}_{\mathbb{X}} &= L_{\mathbb{X}U}\mathcal{F}_U + L_{\mathbb{X}\mathbb{X}}\mathcal{F}_{\mathbb{X}} + L_{\mathbb{X}\mathbb{Y}}\mathcal{F}_{\mathbb{Y}} \\
\mathcal{J}_{\mathbb{Y}} &= L_{\mathbb{Y}U}\mathcal{F}_U + L_{\mathbb{Y}\mathbb{X}}\mathcal{F}_{\mathbb{X}} + L_{\mathbb{Y}\mathbb{Y}}\mathcal{F}_{\mathbb{Y}}.
\end{aligned}$$

Consider the simple case where the thermodynamic forces for mass are zero, that is, $\mathcal{F}_{\mathbb{X}} = \mathcal{F}_{\mathbb{Y}} = 0$. At first, you'd think that would imply that the flow of matter would be zero however there's a cross-term that involves $\mathcal{F}_U$ so the flow for species $\mathbb{X}$ is

$$\begin{aligned}
\mathcal{J}_{\mathbb{X}} &= L_{\mathbb{X}U}\mathcal{F}_U = L_{\mathbb{X}U}\left(\frac{1}{T_R} - \frac{1}{T_L}\right) \\
&\approx -L_{\mathbb{X}U}\left(\frac{T_R - T_L}{T^2}\right) = -\frac{L_{\mathbb{X}U}}{T^2}(T_R - T_L) \qquad (\mathcal{F}_{\mathbb{X}} = \mathcal{F}_{\mathbb{Y}} = 0)
\end{aligned}$$

and similarly for $\mathcal{J}_{\mathbb{Y}}$. This means that a temperature difference can induce a mass flux; this phenomenon is known as *thermophoresis* or the *Soret effect*. While the coefficients L_{UU}, $L_{\mathbb{X}\mathbb{X}}$ and $L_{\mathbb{Y}\mathbb{Y}}$ are required by Second Law to be positive the cross-term $L_{\mathbb{X}U}$ can be positive or negative. In a gas mixture the species with large molar mass typically migrates towards the cold side while the species with small molar mass migrates towards the hot side. Similarly, even when there's no temperature difference so $\mathcal{F}_U = 0$ we can still have energy flow produced by density differences in the species, that is, if $\mathcal{F}_X \neq 0$ or $\mathcal{F}_Y \neq 0$ then $\mathcal{J}_U \neq 0$. This is known as the *Dufour effect* and it depends on the Onsager coefficients L_{UX} and L_{UY}, which can be positive or negative.

Onsager Reciprocity

We see that there are two types of Onsager coefficients. The "proper coefficients", $L_{i,i}$, are the ones that connect a flow $\mathcal{J}_i$ with its corresponding force $\mathcal{F}_i$ (e.g., heat flow with temperature gradient). The "cross-coefficients", $L_{i,j}$, connect a flow $\mathcal{J}_i$ with the other forces $\mathcal{F}_j$, as with the Seebeck and Peltier effects.

Onsager proved a fundamental result regarding these cross-coefficients, namely

$$L_{i,j} = L_{j,i}.$$

This result is known as *Onsager reciprocity* and it is derived from the requirement of microscopic reversibility for systems near equilibrium. Since $L_{UQ} = L_{QU}$ this means that the Seebeck and Peltier effects are essentially two sides of the same coin, that is, the underlying physics is the same for all cross-coefficient effects. Similarly, the Onsager coefficients for the Soret and Dufour effects are equal (e.g., $L_{XU} = L_{UX}$).

* * *

In classical mechanics, Lagrangian dynamics and Hamiltonian dynamics have many similarities yet both are useful in their own fashion. The next chapter describes thermodynamic potentials, which are alternatives to using $U(S, V, \mathbf{M})$. You briefly met one of them, enthalpy, back in Chapter 4, and now it's time to introduce you to the whole family.

Chapter 12

Thermodynamic Potentials

This chapter introduces the concept of thermodynamic potentials and yet we've already been using two of them: internal energy, U, and enthalpy, H. This chapter adds two more: Helmholtz free energy, F, and Gibbs free energy, G. We'll discuss their importances and use them to derive a few thermodynamic identities (and even more relations in the next chapter).

12.1 Internal Energy

Many fundamental physical laws are based on an extremum principle. For example, geometric optics is based on the principle of least time (Fermat's principle); Lagrangian mechanics is based on the principle of least action (Maupertuis's principle). In thermodynamics we have established that for an isolated system (fixed U and V with $d_e\mathbf{M} = 0$), the entropy S is maximum at equilibrium. We now establish a related result: for a closed system ($d_e\mathbf{M} = 0$) with fixed S and V the internal energy is *minimum* at thermodynamic equilibrium.

Consider the heterogeneous closed system illustrated in Figure 12.1. Initially the Left subsystem is at a higher temperature than the Right so the thermodynamic force $\mathcal{F}_U > 0$ and energy flows towards the right ($\mathcal{J}_U > 0$). By the Second Law $d_i S/dt > 0$ but this system has an "entropy overflow valve" which keeps S constant by having $d_e S/dt = -d_i S/dt$. Physically this means that heat leaves the system as we approach equilibrium.

Let's show that internal energy is *minimum* at thermodynamic equilibrium. From the First Law, $dU = dQ + dW$ and volume is fixed so $dW = -pdV = 0$. Since $dQ = Td_e S$ for $d_e\mathbf{M} = 0$ then

$$\frac{dU}{dt} = T\frac{d_e S}{dt} = -T\frac{d_i S}{dt} \leq 0 \qquad (\text{Fixed } S, V;\ d_e\mathbf{M} = 0).$$

So with these (admittedly unusual) constraints the internal energy always

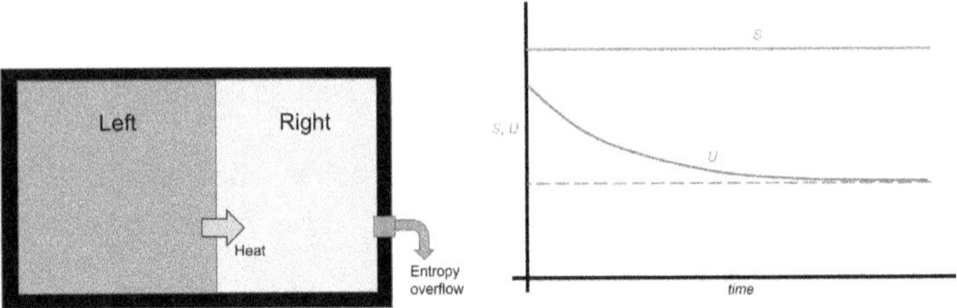

Figure 12.1: Heterogeneous closed system with fixed S and V. Total entropy is held constant using an "entropy overflow valve."

decreases until we reach thermodynamic equilibrium, at which point $d_i S/dt = 0$.

It was stated in Chapter 7 that thermodynamics could be formulated either in the entropy representation, defining $S(U, V, \mathbf{M})$, or in the energy representation, defining $U(S, V, \mathbf{M})$. The Second Law established a maximum principle for entropy and we've now established a corresponding minimum principle for internal energy. But, as you're about to see, there are other thermodynamic potentials with similar minimum principles.

12.2　Helmholtz Free Energy

The Helmholtz free energy is defined as

$$F = U - TS.$$

Note that the Helmholtz free energy is sometimes called simply the "free energy", which is why the symbol F is used.* In some textbooks the symbol A (for "Arbeit", which is work in German) is used instead of F.

For a closed system with fixed T and V the Helmholtz free energy is minimum at equilibrium; here's the proof: From the definition of F, at fixed T

$$dF = dU - TdS = dU - T(d_e S + d_i S) \qquad \text{(Fixed } T\text{)}.$$

Using First Law, $dU = dQ - pdV$ so

$$dF = (dQ - pdV) - Td_e S - Td_i S = -pdV - Td_i S \qquad \text{(Fixed } T;\ d_e\mathbf{M} = 0)$$

since $dQ = Td_e S$ for $d_e\mathbf{M} = 0$. Finally, for fixed V

$$dF = -Td_i S \qquad \text{and} \qquad \frac{dF}{dt} = -T\frac{d_i S}{dt} \qquad \text{(Fixed } T, V;\ d_e\mathbf{M} = 0).$$

By the Second Law $d_i S/dt \geq 0$, so the Helmholtz free energy always decreases until we reach thermodynamic equilibrium, at which point $d_i S/dt = 0$. As

---

*The term "free" is in the sense of "available" as in *Are you free tonight?*.

such, F is the thermodynamic potential for closed systems with fixed T and V.

Note that when a thermodynamic force, $\mathcal{F}$, drives a thermodynamic flow, $\mathcal{J}$, in a closed system then

$$\frac{dF}{dt} = -T\mathcal{F}\mathcal{J}$$

when T and V are constant. For example, in a homogeneous system with a chemical reaction

$$\frac{dF}{dt} = -\mathcal{A}\frac{d\xi}{dt} \quad \text{and} \quad \Delta F_{\text{AB}} = -\int_{\text{A}}^{\text{B}} \mathcal{A}(\xi)d\xi$$

where $\mathcal{A}$ and ξ are the chemical affinity and the extent of reaction (see Section 9.3). Since chemical affinity is the net difference in chemical potential between reactants and products this says that the amount of available ("free") chemical energy obtained from the chemical reaction equals the change in the Helmholtz free energy. This is different from the change of internal energy because temperature is fixed so heat possibly enters or leaves the system.

From the definition of Helmholtz free energy $dF = d(U - TS)$ so

$$dF = dU - TdS - SdT.$$

Using the Gibbs equation

$$dU = TdS - pdV + \boldsymbol{\mu} \cdot d\mathbf{M}$$

we have a Gibbs equation for Helmholtz free energy

$$dF = -SdT - pdV + \boldsymbol{\mu} \cdot d\mathbf{M}.$$

Since $F(T, V, \mathbf{M})$ we have

$$\begin{aligned}
dF(T, V, \mathbf{M}) \;=\; & \left(\frac{\partial F}{\partial T}\right)_{V,\mathbf{M}} dT + \left(\frac{\partial F}{\partial V}\right)_{T,\mathbf{M}} dV \\
& + \left(\frac{\partial F}{\partial M_1}\right)_{T,V,M_{j\neq 1}} dM_1 + \text{Similar terms for } dM_2, dM_3, \ldots
\end{aligned}$$

so from the above

$$\left(\frac{\partial F}{\partial T}\right)_{V,\mathbf{M}} = -S \qquad \left(\frac{\partial F}{\partial V}\right)_{T,\mathbf{M}} = -p \qquad \left(\frac{\partial F}{\partial M_i}\right)_{T,V,M_{j\neq i}} = \mu_i.$$

Compare this with the similar result for $U(S, V, \mathbf{M})$

$$\left(\frac{\partial U}{\partial S}\right)_{V,\mathbf{M}} = T \qquad \left(\frac{\partial U}{\partial V}\right)_{S,\mathbf{M}} = -p \qquad \left(\frac{\partial U}{\partial M_i}\right)_{S,V,M_{j\neq i}} = \mu_i.$$

Finally, notice that at constant temperature $(dT = 0)$ in a closed, inert system $(d\mathbf{M} = 0)$ we have $dF = -pdV$ so $\Delta F_{\text{AB}} = W_{\text{AB}} = -\mathcal{W}_{\text{AB}}$, that is, the work done by the system equals the loss of Helmholtz free energy.

12.3   Legendre Transformations*

Clearly $U(S, V, \mathbf{M})$ and $F(T, V, \mathbf{M})$ are closely related and that's because they are an example of a *Legendre transformation*. These types of transformations are used in physics when we'd like to introduce a change of variable for a function but the new variable is related to the function itself.

Suppose you have a function $f(x)$ but you find that the independent variable x is inconvenient to work with. You prefer to use as the independent variable w, defined as

$$w = \frac{df}{dx},$$

which defines the change of variable $w(x) \Leftrightarrow x(w)$. However, if you simply use $f(w) \equiv f(x(w))$ (i.e., replace x with w in f) then you'll lose information.

What you want is to formulate a new function, $g(w)$, that has all the same information as $f(x)$. A simple way to do this is with a mathematically reversible transformation, such as

$$f(x), \quad w = \frac{df}{dx} \qquad \Leftrightarrow \qquad g(w), \quad x = -\frac{dg}{dw}.$$

This can be accomplished by defining the new function as

$$g(w) = f(x(w)) - x(w)w,$$

which is known as a Legendre transformation. Given $f(x)$ you can find $g(w)$ and vice versa.

Example 12.1:   You are doing research in yarn theory, a hot new field of physics. The systems that you study are characterized by the function $f(x) = \mathfrak{A}e^{\mathfrak{a}x}$. Formulate the Legendre transformation that allows you to change variables from x to $w = df/dx$.
Solution: In this example

$$w = \frac{df}{dx} = \mathfrak{a}\mathfrak{A}e^{\mathfrak{a}x} \qquad \text{so} \qquad x = \frac{\ln(w/\mathfrak{a}\mathfrak{A})}{\mathfrak{a}}.$$

First notice that since $f(x) = \mathfrak{A}e^{\mathfrak{a}x}$ then $f(w) = w/\mathfrak{a}$ for all values of $\mathfrak{A}$. Our original expression, $f(x)$, has two parameters, $\mathfrak{A}$ and $\mathfrak{a}$, however $f(w)$ only has $\mathfrak{a}$ so information is lost.

By a Legendre transformation we introduce the function

$$g(w) = f(x(w)) - x(w)w = \mathfrak{A}e^{\mathfrak{a}(\ln(w/\mathfrak{a}\mathfrak{A})/\mathfrak{a})} - \frac{\ln(w/\mathfrak{a}\mathfrak{A})}{\mathfrak{a}}w = \frac{w(1 - \ln(w/\mathfrak{a}\mathfrak{A}))}{\mathfrak{a}}$$

and you can check that

$$-\frac{dg}{dw} = -\frac{(1 - \ln(w/\mathfrak{a}\mathfrak{A})) - 1}{\mathfrak{a}} = x$$

as required. The function $g(w)$ has all the information originally contained in $f(x)$.

Legendre transformations are also used in classical mechanics. For example, for generalized coordinates and momenta q and p the Hamiltonian, $\mathcal{H}(q,p)$, is obtained from the Lagrangian, $\mathcal{L}(q,\dot{q})$, as

$$\mathcal{H} = -(\mathcal{L} - p\dot{q}).$$

This is a Legendre transformation that replaces velocity, $\dot{q}$ with momentum, p. Since momentum is defined as, $p = (\partial\mathcal{L}/\partial\dot{q})_q$ we cannot simply use $\mathcal{L}(q,p)$ directly.

Getting back to thermodynamics, our definition of Helmholtz free energy, $F = U - TS$ is a Legendre transformation that takes $U(S, V, \mathbf{M})$ and replaces S with T to produce $F(T, V, \mathbf{M})$, as illustrated in the next example.

Example 12.2: From our results in Section 5.2 for the entropy of a pure ideal gas, the internal energy may be written as

$$U(S, V, M) = \mathfrak{A}(V, M) \exp\left(\frac{S}{c_V M}\right)$$

where $\mathfrak{A}$ is a complicated function of V and M. Find $F(T, V, M)$.
Solution: The temperature is defined as

$$T = \left(\frac{\partial U}{\partial S}\right)_{V,M}$$

so for this ideal gas

$$T = \mathfrak{a}\mathfrak{A}e^{\mathfrak{a}S} \qquad \text{and} \qquad S = \frac{\ln(T/\mathfrak{a}\mathfrak{A})}{\mathfrak{a}}$$

where $\mathfrak{a} = 1/(c_V M)$. Since $F = U - TS$, from the results of the previous example with $f = U$, $g = F$, $x = S$, and $w = T$ we get the Helmholtz free energy

$$F(T, V, M) = \frac{T(1 - \ln(T/\mathfrak{a}\mathfrak{A}))}{\mathfrak{a}} = c_V MT\{1 - \ln(c_V MT) + \ln \mathfrak{A}(V, M)\}.$$

From this result we may reverse the transformation (i.e., $U = F + TS$) and we know that we'll recover the original expression for the internal energy. Note that while it's possible (and often useful) to formulate $U(T, V, M)$, we do lose information when replacing S with T; in this example $U = T/\mathfrak{a} = c_V MT$ so we lose $\mathfrak{A}(V, M)$. Using $F(T, V, M)$ we can get the equations of state for p and μ but we cannot get them from $U(T, V, M)$.

12.4   Enthalpy and Gibbs Free Energy

Section 4.2 introduced the enthalpy and you can see that this is also a Legendre transformation

$$H(S, p, \mathbf{M}) = U(S, p, \mathbf{M}) + pV(S, p, \mathbf{M})$$

replacing V with $-p$ as the independent variable. Enthalpy is another thermodynamic potential and by a similar derivation as we've already performed above for U and F, you get

$$\frac{dH}{dt} = -T\frac{d_{\mathrm{i}}S}{dt} \leq 0 \qquad (\text{Fixed } S, p; \ d_{\mathrm{e}}\mathbf{M} = 0).$$

As such, H is the thermodynamic potential for closed systems with fixed S and p.

There's one more thermodynamic potential that's worth mentioning: the Gibbs free energy. As with the others, it is the result of a Legendre transformation

$$G(T, p, \mathbf{M}) = H - TS = F + pV = U - TS + pV.$$

Again by a similar derivation

$$\frac{dG}{dt} = -T\frac{d_{\mathrm{i}}S}{dt} \leq 0 \qquad (\text{Fixed } T, p; \ d_{\mathrm{e}}\mathbf{M} = 0).$$

As such, G is the thermodynamic potential for closed systems with fixed T and p.

The Gibbs equations for these thermodynamic potentials are

$$
\begin{aligned}
dH(S, p, \mathbf{M}) &= T\,dS + V\,dp + \boldsymbol{\mu} \cdot d\mathbf{M} \\
dG(T, p, \mathbf{M}) &= -S\,dT + V\,dp + \boldsymbol{\mu} \cdot d\mathbf{M}.
\end{aligned}
$$

From these expressions,

$$\left(\frac{\partial H}{\partial S}\right)_{p,\mathbf{M}} = T \qquad \left(\frac{\partial H}{\partial p}\right)_{S,\mathbf{M}} = V \qquad \left(\frac{\partial H}{\partial M_i}\right)_{S,p,M_{j\neq i}} = \mu_i$$

and

$$\left(\frac{\partial G}{\partial T}\right)_{p,\mathbf{M}} = -S \qquad \left(\frac{\partial G}{\partial p}\right)_{T,\mathbf{M}} = V \qquad \left(\frac{\partial G}{\partial M_i}\right)_{T,p,M_{j\neq i}} = \mu_i.$$

There is an interesting result that comes from this last expression. In the case where the system is a pure substance (i.e., only a single species) we have

$$\left(\frac{\partial G}{\partial M}\right)_{T,p} = \mu.$$

Since all the thermodynamic potentials are extensive functions, $G(T, p, \mathcal{C}M) = \mathcal{C}G(T, p, M)$. This means that

$$G(T, p, M) = M\mu(T, p)$$

so the chemical potential is the Gibbs free energy per unit mass. Is there a similar relation for the other thermodynamic potentials? No, because G is the only one for which the only independent extensive variable is M. By the way, when there's more than one species $G = \boldsymbol{\mu} \cdot \mathbf{M} = \sum_k M_k \mu_k$; we'll derive this in the next chapter.

Example 12.3: Show that the change in the Gibbs free energy for a ideal gas in an isothermal expansion is equal to the amount of work done by the gas. Take the system to be closed and inert.
Solution: In general, $dG = -SdT + Vdp + \boldsymbol{\mu} \cdot d\mathbf{M}$ but for an isothermal expansion in a closed, inert system $dT = 0$ and $d\mathbf{M} = 0$ so in this case $dG = Vdp$. The change in the Gibbs free energy is

$$\Delta G_{\mathrm{AB}} = \int_{\mathrm{A}}^{\mathrm{B}} V(p)dp = NRT \int_{\mathrm{A}}^{\mathrm{B}} \frac{dp}{p} = NRT \ln(p_{\mathrm{B}}/p_{\mathrm{A}}) = -NRT \ln(V_{\mathrm{B}}/V_{\mathrm{A}})$$

since for an ideal gas $V = NRT/p$. The work done by an isothermal expansion of an ideal gas is

$$\mathcal{W}_{\mathrm{AB}} = \int_{\mathrm{A}}^{\mathrm{B}} pdV = NRT \int_{\mathrm{A}}^{\mathrm{B}} \frac{dV}{V} = NRT \ln(V_{\mathrm{B}}/V_{\mathrm{A}})$$

Another way to arrive at the result is to note that for an ideal gas at constant temperature $pdV = -Vdp$ since $d(pV) = d(NRT) = 0$.

12.5 Summary of Thermodynamic Potentials

Before this chapter we had two equivalent formulations of the Gibbs equation, the energy representation using $U(S, V, \mathbf{M})$

$$dU(S, V, \mathbf{M}) = TdS - pdV + \boldsymbol{\mu} \cdot d\mathbf{M} \qquad \text{(Internal Energy)}$$

and the entropy representation using $S(U, V, \mathbf{M})$

$$dS(U, V, \mathbf{M}) = \frac{1}{T}dU + \frac{p}{T}dV - \frac{1}{T}\boldsymbol{\mu} \cdot d\mathbf{M} \qquad \text{(Entropy)}.$$

The Second Law says that entropy of an isolated system (fixed U, V, and $d_e\mathbf{M} = 0$) is maximum at thermodynamic equilibrium. In this chapter we found that there's a corresponding result for internal energy; it's minimum at equilibrium for closed systems ($d_e\mathbf{M} = 0$) with fixed S and V. As such we identify U as a thermodynamic potential.

This result led us to formulate three more thermodynamic potentials by Legendre transformations. The corresponding Gibbs equations are

$$
\begin{aligned}
dF(T,V,\mathbf{M}) &= -SdT - pdV + \boldsymbol{\mu}\cdot d\mathbf{M} && \text{(Helmholtz free energy)}\\
dH(S,p,\mathbf{M}) &= TdS + Vdp + \boldsymbol{\mu}\cdot d\mathbf{M} && \text{(Enthalpy)}\\
dG(T,p,\mathbf{M}) &= -SdT + Vdp + \boldsymbol{\mu}\cdot d\mathbf{M} && \text{(Gibbs free energy)}.
\end{aligned}
$$

Any one of these could be used as the starting point for the formulation of thermodynamics. These are the most common thermodynamic potentials but using Legendre transformations we can define others. For example, the Landau potential, also known as the Grand potential, $\Phi(T,V,\boldsymbol{\mu}) = F - \boldsymbol{\mu}\cdot\mathbf{M}$ is minimum at equilibrium for open systems at fixed T, V, and $\boldsymbol{\mu}$.

Thermodynamic Potentials and Statistical Mechanics

In statistical mechanics the theory of ensembles allows us to formulate thermodynamics from a microscopic perspective. Using the *microcanonical ensemble* we have

$$S(U,V,\mathbf{M}) = k_B \ln \Omega(U,V,\mathbf{M})$$

where Ω is the multiplicity of microscopic states for fixed $(U,V,\mathbf{M})$. Alternatively, using the *canonical ensemble* we have

$$F(T,V,\mathbf{M}) = -k_B T \ln \mathcal{Z}(T,V,\mathbf{M})$$

where $\mathcal{Z}$ is the *partition function*, which is a weighted sum of microscopic states for fixed $(T,V,\mathbf{M})$. The canonical ensemble is usually easier to work with and once we have $F(T,V,\mathbf{M})$ we can obtain expressions for any other thermodynamic quantity. For example, entropy and pressure are

$$S = -\left(\frac{\partial F}{\partial T}\right)_{V,\mathbf{M}} \quad \text{and} \quad p = -\left(\frac{\partial F}{\partial V}\right)_{T,\mathbf{M}}.$$

For some systems (e.g., quantum ideal gases) the *grand canonical ensemble* is preferable. It uses the Landau potential with

$$\Phi(T,V,\boldsymbol{\mu}) = -k_B T \ln \mathcal{Z}(T,V,\boldsymbol{\mu})$$

where $\mathcal{Z}$ is the grand partition function.

Example 12.4: Conduction electrons are often modeled as a pure quantum ideal gas (i.e., free electron model). From the grand canonical ensemble we find that the Landau potential is

$$\Phi(T,V,\mu) = -\frac{2k_B T V}{\lambda_T^3} f_{5/2}(z)$$

where $\lambda_T = \sqrt{2\pi\hbar^2/mk_B T}$ is the thermal wavelength, $\mathfrak{z} = \exp(m\mu/k_B T)$ is the fugacity, and $f_{5/2}$ is a Fermi integral. Find the pressure in terms of T, V, and μ.

Solution: From the definition of the Landau potential the corresponding Gibbs equation is

$$d\Phi(T, V, \mu) = -S dT - p dV - M d\mu.$$

The pressure in this gas is

$$p = -\left(\frac{\partial \Phi}{\partial V}\right)_{T,\mu} = \frac{2k_B T}{\lambda_T^3} f_{5/2}(\mathfrak{z})$$

since $\mathfrak{z}(T, \mu)$ is constant for fixed T and μ. Other thermodynamic properties, such as density, are obtained by similar manipulations.

$$* * *$$

In the next chapter we'll use our newly formulated thermodynamic potentials to derive a variety of important thermodynamic relations. These highlight the predictive power of thermodynamics, allowing us to determine properties that are difficult to measure by relating them to others that are easily accessible.

Chapter 13

Thermodynamic Relations

The previous chapter introduced the most common thermodynamic potentials. In this chapter we use them to derive a variety of useful thermodynamic relations. These results allow us to obtain quantities of interest that are difficult to measure experimentally by relating them to other properties that are easily obtained in the lab or by computer simulation.

13.1 Gibbs-Duhem Equation

The thermodynamic potentials are useful in formulating general relations among state variables. We'll start by using the internal energy, $U(S, V, \mathbf{M})$, to derive the *Gibbs Duhem equation*,

$$S dT - V dp + \mathbf{M} \cdot d\boldsymbol{\mu} = 0.$$

Notice the similarity with the Gibbs equation,

$$T dS - p dV + \boldsymbol{\mu} \cdot d\mathbf{M} = dU.$$

The Gibbs-Duhem equation is a consequence of the internal energy being an extensive function.

To derive this result we start with the Euler theorem for extensive functions, which states that if f is an extensive function of extensive variables x_i, that is, if $f(\mathcal{C}x_1, \ldots, \mathcal{C}x_n) = \mathcal{C}f(x_1, \ldots, x_n)$ then

$$f(x_1, \ldots, x_n) = x_1 \left(\frac{\partial f}{\partial x_1}\right)_{x_j \neq x_1} + \ldots + x_n \left(\frac{\partial f}{\partial x_n}\right)_{x_j \neq x_n} = \sum_{i=1}^{n} x_i \left(\frac{\partial f}{\partial x_i}\right)_{x_j \neq x_i}.$$

Using this theorem

$$U(S, V, \mathbf{M}) = S \left(\frac{\partial U}{\partial S}\right)_{V, \mathbf{M}} + V \left(\frac{\partial U}{\partial V}\right)_{S, \mathbf{M}} + \sum_{k=1}^{K} M_k \left(\frac{\partial U}{\partial M_k}\right)_{S, V, M_{j \neq k}}.$$

Yet from the Gibbs equation in the energy representation

$$T = \left(\frac{\partial U}{\partial S}\right)_{V,\mathbf{M}} \qquad p = -\left(\frac{\partial U}{\partial V}\right)_{S,\mathbf{M}} \qquad \mu_k = \left(\frac{\partial U}{\partial M_k}\right)_{S,V,M_{j\neq k}}$$

so the internal energy is,

$$U = TS - pV + \boldsymbol{\mu} \cdot \mathbf{M}$$

which is an example of an *Euler relation*. The differential of this expression is

$$
\begin{aligned}
dU \;\; &= \;\; TdS + SdT - pdV - Vdp + \sum_{k=1}^{K}\mu_k dM_k + \sum_{k=1}^{K} M_k d\mu_k \\
&= \;\; TdS - pdV + \sum_{k=1}^{K}\mu_k dM_k + \left\{ SdT - Vdp + \sum_{k=1}^{K} M_k d\mu_k \right\}.
\end{aligned}
$$

From the Gibbs equation we see that the term in the curly braces is zero, which is the Gibbs-Duhem equation. Note that we don't have a similar relation coming from the other thermodynamic potentials since $U(S,V,\mathbf{M})$ is the only one for which all of the independent variables are extensive variables.

Example 13.1: Show that $G(T,p,\mathbf{M}) = \boldsymbol{\mu} \cdot \mathbf{M} = \sum_k \mu_k M_k$.
Solution: Recall from Section 12.4 that $dG(T,p,\mathbf{M}) = -SdT + Vdp + \boldsymbol{\mu} \cdot d\mathbf{M}$. Applying the Gibbs-Duhem equation gives

$$dG(T,p,\mathbf{M}) = d\mathbf{M} \cdot \boldsymbol{\mu} + \boldsymbol{\mu} \cdot d\mathbf{M} = d(\boldsymbol{\mu} \cdot \mathbf{M}),$$

which gives the desired result. We can also derive this result using our Euler relation for internal energy, $U = TS - pV + \boldsymbol{\mu} \cdot \mathbf{M}$, and the definition of Gibbs free energy, $G = U - TS + pV$. By the way, since $G = F + pV$ the Landau potential is $\Phi = F - \boldsymbol{\mu} \cdot \mathbf{M} = -pV$.

13.2 Maxwell Relations

When you first met the Gibbs equation back in Section 7.1 we saw that

$$T = \left(\frac{\partial U}{\partial S}\right)_{V,\mathbf{M}} \qquad \text{and} \qquad p = -\left(\frac{\partial U}{\partial V}\right)_{S,\mathbf{M}}.$$

Using the partial derivative identity from Section 2.2

$$\left(\frac{\partial}{\partial x}\right)_y \left(\frac{\partial f}{\partial y}\right)_x = \left(\frac{\partial}{\partial y}\right)_x \left(\frac{\partial f}{\partial x}\right)_y = \frac{\partial^2 f}{\partial x \partial y} \qquad\qquad \text{(D4)}$$

we have

$$\frac{\partial^2 U}{\partial S \partial V} = \left(\frac{\partial}{\partial V}\right)_{S,\mathbf{M}} \left(\frac{\partial U}{\partial S}\right)_{V,\mathbf{M}} = \left(\frac{\partial}{\partial S}\right)_{V,\mathbf{M}} \left(\frac{\partial U}{\partial V}\right)_{S,\mathbf{M}}$$

or

$$\left(\frac{\partial T}{\partial V}\right)_{S,\mathbf{M}} = - \left(\frac{\partial p}{\partial S}\right)_{V,\mathbf{M}} \qquad \text{(M1)}.$$

This identity is called a *Maxwell relation*; it is a relation between $T, V, p,$ and S that comes from the fact that dU is an exact differential.

There are three other common Maxwell relations that come from the three other basic thermodynamic potentials, H, F, and G:

$$\left(\frac{\partial T}{\partial p}\right)_{S,\mathbf{M}} = \left(\frac{\partial V}{\partial S}\right)_{p,\mathbf{M}} \qquad \text{(M2)}$$

$$\left(\frac{\partial S}{\partial V}\right)_{T,\mathbf{M}} = \left(\frac{\partial p}{\partial T}\right)_{V,\mathbf{M}} \qquad \text{(M3)}$$

$$\left(\frac{\partial S}{\partial p}\right)_{T,\mathbf{M}} = - \left(\frac{\partial V}{\partial T}\right)_{p,\mathbf{M}} \qquad \text{(M4)}.$$

These are the most common Maxwell relations but there are others that can be derived involving derivatives with μ_k and M_k (see Table 13.1).

	$dU = TdS - pdV + \sum_k \mu_k dM_k$	$dF = -SdT - pdV + \sum_k \mu_k dM_k$	
(M1)	$\left(\dfrac{\partial T}{\partial V}\right)_{S,\mathbf{M}} = - \left(\dfrac{\partial p}{\partial S}\right)_{V,\mathbf{M}}$	$\left(\dfrac{\partial S}{\partial V}\right)_{T,\mathbf{M}} = \left(\dfrac{\partial p}{\partial T}\right)_{V,\mathbf{M}}$	(M3)
(M1')	$\left(\dfrac{\partial T}{\partial M_k}\right)_{S,V,M_i \neq M_k} = \left(\dfrac{\partial \mu_k}{\partial S}\right)_{V,\mathbf{M}}$	$\left(\dfrac{\partial S}{\partial M_k}\right)_{T,V,M_i \neq M_k} = - \left(\dfrac{\partial \mu_k}{\partial T}\right)_{V,\mathbf{M}}$	(M3')
(M1")	$\left(\dfrac{\partial p}{\partial M_k}\right)_{S,V,M_i \neq M_k} = - \left(\dfrac{\partial \mu_k}{\partial V}\right)_{S,\mathbf{M}}$	$\left(\dfrac{\partial p}{\partial M_k}\right)_{T,V,M_i \neq M_k} = - \left(\dfrac{\partial \mu_k}{\partial V}\right)_{T,\mathbf{M}}$	(M3")
	$dH = TdS + Vdp + \sum_k \mu_k dM_k$	$dG = -SdT + Vdp + \sum_k \mu_k dM_k$	
(M2)	$\left(\dfrac{\partial T}{\partial p}\right)_{S,\mathbf{M}} = \left(\dfrac{\partial V}{\partial S}\right)_{p,\mathbf{M}}$	$\left(\dfrac{\partial S}{\partial p}\right)_{T,\mathbf{M}} = - \left(\dfrac{\partial V}{\partial T}\right)_{p,\mathbf{M}}$	(M4)
(M2')	$\left(\dfrac{\partial T}{\partial M_k}\right)_{S,p,M_i \neq M_k} = \left(\dfrac{\partial \mu_k}{\partial S}\right)_{p,\mathbf{M}}$	$\left(\dfrac{\partial S}{\partial M_k}\right)_{T,p,M_i \neq M_k} = - \left(\dfrac{\partial \mu_k}{\partial T}\right)_{p,\mathbf{M}}$	(M4')
(M2")	$\left(\dfrac{\partial V}{\partial M_k}\right)_{S,p,M_i \neq M_k} = \left(\dfrac{\partial \mu_k}{\partial p}\right)_{S,\mathbf{M}}$	$\left(\dfrac{\partial V}{\partial M_k}\right)_{T,p,M_i \neq M_k} = \left(\dfrac{\partial \mu_k}{\partial p}\right)_{T,\mathbf{M}}$	(M4")

Table 13.1: Table of Maxwell relations obtained from U, H, F, and G.

Example 13.2: Use internal energy to derive a Maxwell relation for $(\partial \mu_k / \partial V)_{S,\mathbf{M}}$.

Solution: We start by writing the Gibbs equation for $dU(S, V, \mathbf{M})$ as

$$dU = TdS - pdV + \mu_k dM_k + \sum_{j=1, j \neq k}^{K} \mu_j dM_j$$

where the sum is over all species *except* for k. For fixed S and $M_{j\neq k}$ this is just

$$dU(V, M_k) = -pdV + \mu_k dM_k \qquad (\text{Fixed } S, M_{j\neq k})$$

so

$$p = -\left(\frac{\partial U}{\partial V}\right)_{S,M_{j\neq k},M_k} = -\left(\frac{\partial U}{\partial V}\right)_{S,\mathbf{M}} \quad \text{and} \quad \mu_k = \left(\frac{\partial U}{\partial M_k}\right)_{S,V,M_{j\neq k}}.$$

Again, using the fact that

$$\frac{\partial^2 U}{\partial V \partial M_k} = \left(\frac{\partial}{\partial V}\right)_{S,\mathbf{M}}\left(\frac{\partial U}{\partial M_k}\right)_{S,V,M_{j\neq k}} = \left(\frac{\partial}{\partial M_k}\right)_{S,V,M_{j\neq k}}\left(\frac{\partial U}{\partial V}\right)_{S,\mathbf{M}}$$

gives the Maxwell relation

$$\left(\frac{\partial \mu_k}{\partial V}\right)_{S,\mathbf{M}} = -\left(\frac{\partial p}{\partial M_k}\right)_{S,V,M_{j\neq k}}.$$

This is listed as (M1") in Table 13.1.

Each Maxwell relation tells an interesting story. For example,

$$\left(\frac{\partial S}{\partial V}\right)_{T,\mathbf{M}} = \left(\frac{\partial p}{\partial T}\right)_{V,\mathbf{M}} \qquad (\text{M3})$$

says that for all materials the rate at which entropy increases in an isothermal expansion is always equal to the rate at which pressure increases as we increase temperature while holding the material at constant volume. Would you have guessed that this was true? The Maxwell relations are a prime example of the predictive power of thermodynamics.

13.3 Applying Thermodynamic Relations

The Gibbs-Duhem equation is useful in a variety of derivations. For example, it tells us that T, p, and μ are not independent variables. For a pure substance one usually takes p and T as independent intensive variables so $\mu(T,p)$ with

$$d\mu(T,p) = -\frac{S}{M}dT + \frac{V}{M}dp = \frac{1}{\rho}\left(-sdT + dp\right)$$

where $s = S/V$ and $\rho = M/V$ are the entropy density and mass density.

Example 13.3: For a pure ideal gas in a gravitational field (acceleration g) the mass density decreases with height, z. At constant temperature,

$$\rho(z) = \rho(0)\exp(-mgz/k_\mathrm{B}T).$$

Find the dependance on height of the chemical potential.
Solution: The Gibbs-Duhem equation at constant temperature is

$$d\mu(T, p(z)) = \frac{dp}{\rho} = \frac{k_B T}{m} \frac{d\rho}{\rho}$$

using the ideal gas law written as $p = \rho k_B T / m$. Integrating gives

$$\mu(T, p(z)) - \mu(T, p(0)) = \frac{k_B T}{m} \left[\ln(\rho(z)) - \ln(\rho(0)) \right] = -gz$$

so chemical potential decreases with height. Finally, it's useful to define a function $\mu^* = \mu + gz$, which is constant with height. There is an analogous function $\mu^* = \mu + q\phi/m$, called the *electrochemical potential*, defined for particles of charge q in an electric potential, ϕ.

Maxwell relations are useful in linking material properties that are easily measured experimentally to other quantities of interest that are not readily accessible. Three properties that are commonly measured are: specific heat capacity at constant pressure, c_p, the coefficient of thermal expansion, α, and the isothermal compressibility, κ_T. Various relations among different quantities can be derived from these three using Maxwell relations.

Example 13.4: Show that, for fixed **M**,

$$c_V = c_p - \frac{T\alpha^2}{\rho\kappa_T}.$$

This is a useful result since c_p is usually easier to measure than c_V. Also find $c_p - c_V$ for iron at $20°\text{C}$.
Solution: Using the identity

$$\left(\frac{\partial f}{\partial x}\right)_z = \left(\frac{\partial f}{\partial x}\right)_y + \left(\frac{\partial f}{\partial y}\right)_x \left(\frac{\partial y}{\partial x}\right)_z \qquad \text{(D5)}$$

we have

$$\left(\frac{\partial S}{\partial T}\right)_{p,\mathbf{M}} = \left(\frac{\partial S}{\partial T}\right)_{V,\mathbf{M}} + \left(\frac{\partial S}{\partial V}\right)_{T,\mathbf{M}} \left(\frac{\partial V}{\partial T}\right)_{p,\mathbf{M}}$$

Since $C_V = T(\partial S/\partial T)_{V,\mathbf{M}}$ and $C_p = T(\partial S/\partial T)_{p,\mathbf{M}}$, the first two terms are

$$\left(\frac{\partial S}{\partial T}\right)_{p,\mathbf{M}} = \frac{C_p}{T} \qquad \text{and} \qquad \left(\frac{\partial S}{\partial T}\right)_{V,\mathbf{M}} = \frac{C_V}{T}.$$

Transforming with a Maxwell relation and then applying the cyclic chain rule, the next term is

$$\left(\frac{\partial S}{\partial V}\right)_{T,\mathbf{M}} = \left(\frac{\partial p}{\partial T}\right)_{V,\mathbf{M}} = -\frac{(\partial V/\partial T)_{p,\mathbf{M}}}{(\partial V/\partial p)_{T,\mathbf{M}}} = \frac{\alpha V}{\kappa_T V}.$$

The last term is just $V\alpha$ so assembling the pieces

$$\frac{C_p}{T} = \frac{C_V}{T} + \frac{\alpha}{\kappa_T}(V\alpha)$$

which is the desired result after rearranging the terms and dividing through by M. For iron,

$$c_p - c_V = \frac{T\alpha^2}{\rho\kappa_T} = \frac{(273\ \text{K})(3.5 \times 10^{-5}\ \text{K}^{-1})^2}{(7900\ \text{kg/m}^2)(5.9 \times 10^{-12}\ \text{Pa}^{-1})} = 7.2\ \text{J/kg K}.$$

By comparison $c_p = 450\ \text{J/kg K}$.

Example 13.5: Express the rate of change of temperature when volume changes adiabatically in terms of α, κ_T, and C_V. Take the system to be closed and inert (i.e., $\mathbf{M}$ is fixed).

Solution: The quantity we want to find is

$$\left(\frac{\partial T}{\partial V}\right)_{S,\mathbf{M}} = -\left(\frac{\partial p}{\partial S}\right)_{V,\mathbf{M}} \qquad \text{(using M1)}$$

$$= -\left(\frac{\partial p}{\partial T}\right)_{V,\mathbf{M}} \Big/ \left(\frac{\partial S}{\partial T}\right)_{V,\mathbf{M}} \qquad \text{(using D1,D2)}.$$

Using (D3′), and the definitions of α, κ_T, and C_V,

$$-\left(\frac{\partial p}{\partial T}\right)_{V,\mathbf{M}} \Big/ \left(\frac{\partial S}{\partial T}\right)_{V,\mathbf{M}} = \frac{\left(\frac{\partial p}{\partial V}\right)_{T,\mathbf{M}}\left(\frac{\partial V}{\partial T}\right)_{p,\mathbf{M}}}{\frac{1}{T}\left(T\left(\frac{\partial S}{\partial T}\right)_{V,\mathbf{M}}\right)} = \frac{\left(\frac{-1}{V\kappa_T}\right)(\alpha V)}{\frac{1}{T}C_V}$$

so

$$\left(\frac{\partial T}{\partial V}\right)_{S,\mathbf{M}} = -\frac{\alpha T}{\kappa_T C_V}.$$

This tells us that if $\alpha > 0$ then the temperature decreases when the system expands by an adiabatic process. On the other hand, if $\alpha < 0$ then an adiabatic expansion causes a rise in temperature. From the stability conditions of the Second Law we know that κ_T and C_V are always positive but α, though usually also positive, is in some cases negative.

Useful relations for thermodynamic potentials can be derived with the aid of Maxwell relations. For example, for fixed $\mathbf{M}$ the Gibbs equation is $dU = TdS - pdV$ so

$$\left(\frac{\partial U}{\partial V}\right)_{T,\mathbf{M}} = T\left(\frac{\partial S}{\partial V}\right)_{T,\mathbf{M}} - p$$

and by the Maxwell relation (M3),

$$\left(\frac{\partial U}{\partial V}\right)_{T,M} = T\left(\frac{\partial p}{\partial T}\right)_{V,M} - p = T^2\left(\frac{\partial}{\partial T}\frac{p}{T}\right)_{V,M}.$$

This result is useful since it complements the expression for the temperature dependance of U,

$$\left(\frac{\partial U}{\partial T}\right)_{V,M} = \left(\frac{dQ}{dT}\right)_{V,M} = C_V$$

as obtained from the First Law. We see that by measuring the heat capacity, $C_V(T,V)$ and the pressure, $p(T,V)$, we can determine the internal energy. That is, the heat capacity tells us how $U(T,V)$ varies with T and the pressure can tell us how $U(T,V)$ varies with V.

Example 13.6: Consider a pure closed system with,

$$(p + f(V,N))(V - bN) = NRT$$

where b is a constant. In general $c_V(V,T,N)$ but show that for this system the specific heat capacity is independent of volume.
Solution: The result we want to derive is that $(\partial c_V/\partial V)_{T,M} = 0$; M is fixed so N is also fixed. Since $c_V = C_V/M$ and $C_V = (\partial U/\partial T)_{V,M}$, we can write

$$\left(\frac{\partial}{\partial V}\right)_{T,M} C_V = \left(\frac{\partial}{\partial V}\right)_{T,M}\left(\frac{\partial U}{\partial T}\right)_{V,M} = \left(\frac{\partial}{\partial T}\right)_{V,M}\left(\frac{\partial}{\partial V}\right)_{T,M} U$$

$$= \left(\frac{\partial}{\partial T}\right)_{V,M}\left\{T^2\left(\frac{\partial}{\partial T}\frac{p}{T}\right)_{V,M}\right\}.$$

Inserting the given equation for $p(T,V,N)$

$$\left(\frac{\partial}{\partial V}\right)_{T,M} C_V = \left(\frac{\partial}{\partial T}\right)_{V,M}\left\{T^2\left(\frac{\partial}{\partial T}\left[\frac{NR}{V - bN} - \frac{f(V,N)}{T}\right]\right)_{V,M}\right\}$$

$$= \left(\frac{\partial}{\partial T}\right)_{V,M} f(V,N) = 0,$$

which is the desired result. Note that for $f(V,N) = a(N/V)^2$ this is the van der Waals gas.

Example 13.7: Show that $U = C_V T - aN^2/V$ for the van der Waals gas.
Solution: Using

$$\left(\frac{\partial U}{\partial V}\right)_{T,M} = T^2\left(\frac{\partial}{\partial T}\frac{p}{T}\right)_{V,M} = T^2\left(\frac{\partial}{\partial T}\left[\frac{NR}{V - Nb} - \frac{a}{T}\left(\frac{N}{V}\right)^2\right]\right)_{V,N}$$

$$= a\left(\frac{N}{V}\right)^2.$$

Also $(\partial U/\partial T)_{V,\mathbf{M}} = C_V$ so we get the desired result by integration since the heat capacity is independent of volume (see previous example).

* * *

You've probably heard that entropy indicates a level of disorder or randomness in a system. Yet up to now there's been no random elements in our formulation of thermodynamics. That's because we've only considered macroscopic systems; the molecules in a glass of water are in constant, random motion yet to us the water appears to be at rest.

In the next and final chapter we extend modern thermodynamics to microscopic systems by adding a link between entropy and probability. Not only will this expand the predictive powers of thermodynamics, it is also a gateway to statistical mechanics.

Chapter 14

Thermodynamic Fluctuations*

Turn back to Section 7.3 and recall the heterogeneous, isolated system illustrated in Figure 7.1. The left and right sides of the system can exchange energy, but not mass or volume. This heat flow results in a change in the total entropy

$$dS = \left(\frac{1}{T_R} - \frac{1}{T_L} \right) dU_R.$$

By the Second Law $dS/dt \geq 0$ so if $T_L > T_R$ then $dU_R/dt > 0$. In words, when the left side is hotter than the right side then energy, as heat, is transferred from Left to Right. Total entropy increases until, at equilibrium, $T_L = T_R = T_{eq}$ and heat flow stops.

This basic picture is accurate at the macroscopic scale but thermodynamics is more complicated at the microscopic scale. We know that materials are composed of molecules in random motion resulting in small variations of U_R, even when the system is at equilibrium. This observation indicates that our interpretation of the Second Law needs to be refined to account for microscopic fluctuations.

14.1   Fluctuation Probabilities

The heterogeneous, isolated system analyzed in Section 7.3 has constant volume and mass so we write the internal energy as

$$U = U_L + U_R = c_V M_L T_L + c_V M_R T_R.$$

For simplicity we've taken c_V to be the same for Left and Right. The total internal energy is constant so $U_L = U - U_R$ and we can write the total entropy

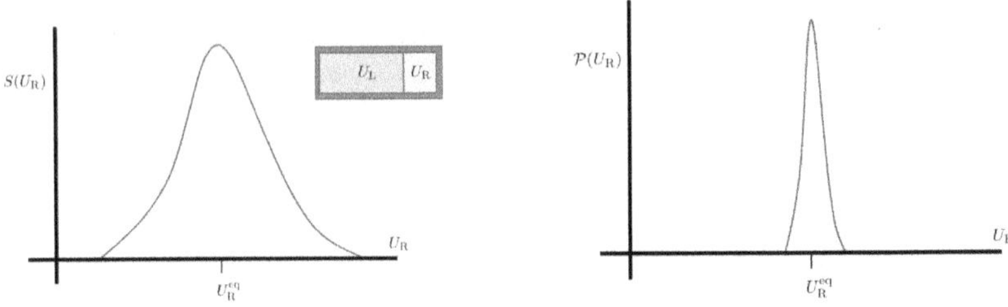

Figure 14.1: Entropy and probability for a heterogeneous, isolated system as functions of U_R, the internal energy of the right side of the system. Total internal energy, $U_L + U_R$, is fixed.

as

$$S(U_R) = S_L(U - U_R) + S_R(U_R).$$

This total entropy is maximum at thermodynamic equilibrium so a graph of $S(U_R)$ should look like Figure 14.1 with a maximum at $U_R^{eq} = c_V M_R T_{eq}$.

To go further we need a relation between entropy and the probability of a fluctuation. We'll use the following ansatz introduced by Einstein: The probability of a state variable being in the range $(x, x + dx)$ is

$$\mathcal{P}(x)\, dx = \mathcal{C} \exp(-(S^{eq} - S(x))/k_B)\, dx$$

where $\mathcal{C}$ is a normalization constant set by the condition $\int \mathcal{P}(x) dx = 1$. As expected, the most probable state is the equilibrium state since $S^{eq} \geq S(x)$ yet due to random fluctuations the value of x will vary, even at equilibrium.

In our example, to find $\mathcal{P}(U_R)$ we need an expression for $S(U_R)$. Using our earlier result

$$\int_{S(U_R)}^{S^{eq}} dS = \int_{U_R}^{U_R^{eq}} \left(\frac{1}{T_R} - \frac{1}{T_L} \right) dU_R.$$

To simplify the calculation we'll assume $M_L \gg M_R$, that is, the left side is treated as a reservoir so $T_L = T_{eq}$. Integrating,

$$S^{eq} - S(U_R) = -c_V M_R \ln(U_R/U_R^{eq}) - \frac{U_R^{eq} - U_R}{T_{eq}}.$$

since $T_R = U_R/c_V M_R$. Taylor expanding the logarithm term gives

$$S^{eq} - S(U_R) \approx \frac{1}{2} c_V M_R \left(\frac{U_R^{eq} - U_R}{U_R^{eq}} \right)^2 = \frac{(U_R^{eq} - U_R)^2}{2 c_V M_R T_{eq}^2}$$

after dropping higher order terms. This quadratic approximation is only accurate near U_R^{eq} but that's good enough since $\mathcal{P}(U_R)$ is sharply peaked (see Figure 14.1).

The probability that the right side of the system has energy U_R is

$$P(U_\mathrm{R})\,dU_\mathrm{R} = \mathcal{C}\exp\left(-\frac{(U_\mathrm{R}^{\mathrm{eq}} - U_\mathrm{R})^2}{2\mathrm{k_B}c_V M_\mathrm{R} T_{\mathrm{eq}}^2}\right)\,dU_\mathrm{R}$$

where $\mathcal{C}$ is a normalization constant. If you've studied probability you may recognize this as the Gaussian function,

$$\mathcal{G}(x; x_0, \sigma) = \frac{1}{\sqrt{2\pi\sigma^2}}\exp\left(-\frac{(x_0 - x)^2}{2\sigma^2}\right).$$

The parameters x_0 and σ are the mean (i.e., average value) and the standard deviation, that is,

$$x_0 \;=\; \langle x\rangle = \int_{-\infty}^{\infty} x\,P(x)dx$$

$$\sigma^2 \;=\; \langle(x - \langle x\rangle)^2\rangle = \int_{-\infty}^{\infty}(x - \langle x\rangle)^2\,P(x)dx.$$

The standard deviation is the square root of the variance so $\sigma = \sqrt{\langle\delta x^2\rangle}$ where $\delta x = x - \langle x\rangle$ is the fluctuation from the mean.

For our example, the mean is $\langle U_\mathrm{R}\rangle = U_\mathrm{R}^{\mathrm{eq}}$ and the standard deviation is

$$\sigma = \sqrt{\langle\delta U_\mathrm{R}^2\rangle} = \sqrt{\langle(U_\mathrm{R} - U_\mathrm{R}^{\mathrm{eq}})^2\rangle} = \sqrt{\mathrm{k_B}c_V M_\mathrm{R}}\;T_{\mathrm{eq}}.$$

The width of the peak of $P(U_\mathrm{R})$ is proportional to σ and the relative magnitude of the standard deviation is

$$\frac{\sigma}{\langle U_\mathrm{R}\rangle} = \frac{\sqrt{\mathrm{k_B}c_V M_\mathrm{R}}\;T_{\mathrm{eq}}}{c_V M_\mathrm{R}T_{\mathrm{eq}}} = \sqrt{\frac{\mathrm{k_B}}{c_V M_\mathrm{R}}}.$$

Typically this relative magnitude is very small for macroscopic systems, for example for a monatomic gas ($c_V = \frac{3}{2}\mathrm{k_B}/m$) it is roughly $1/\sqrt{\mathcal{N}}$ where $\mathcal{N}$ is the number of molecules in the system. The point is that the probability distribution is *very* sharply peaked at the equilibrium value for macroscopic systems.

Example 14.1: What is the probability that in a glass of water a volume element of one cubic micron exceeds the equilibrium temperature of $T_{\mathrm{eq}} = 20\ ^\circ\mathrm{C}$ by more than one degree? Note that this volume element contains roughly 3×10^{10} water molecules.

Solution: The specific heat capacity of water is 1 cal/(K g) so the heat capacity of the volume element is $c_V M = 4.18 \times 10^{-12}$ J/K. The average value of the internal energy in that volume element is $\langle U\rangle = U^{\mathrm{eq}} = c_V M T_{\mathrm{eq}} = 1.22 \times 10^{-9}$ J and the standard deviation is $\sigma = \sqrt{\mathrm{k_B}c_V M}\;T_{\mathrm{eq}} = 2.22 \times 10^{-15}$ J.

The probability that $U \geq U'$ is

$$\mathcal{P}[U \geq U'] \equiv \int_{U'}^{\infty} \mathcal{P}(U)\, dU = \frac{1}{\sqrt{2\pi\sigma^2}} \int_{U'}^{\infty} \exp\left(-\frac{(U^{\text{eq}} - U)^2}{2\sigma^2}\right) dU$$

where $U' = U^{\text{eq}} + (4.18 \times 10^{-12}\ \text{J})$ is the internal energy at 21 °C. This integral cannot be solved in closed form but with the change of variable $z = (U - U^{\text{eq}})/\sqrt{2}\sigma$

$$\mathcal{P}[U \geq U'] = \frac{1}{\sqrt{2\pi\sigma^2}} \int_{z'}^{\infty} \exp\left(-z^2\right)\ (\sqrt{2}\sigma\, dz) = \frac{1}{2}\text{erfc}(z')$$

where $z' = (U' - U^{\text{eq}})/\sqrt{2}\sigma$ and

$$\text{erfc}(x) = \frac{2}{\sqrt{\pi}} \int_{x}^{\infty} \exp\left(-t^2\right) dt$$

is the complementary error function. For our example $z' = 1330$ and the probability is $\mathcal{P}[U \geq U'] \approx 10^{-768000}$, which is *very* small! If the volume element is one million times smaller (roughly 3×10^4 water molecules) then $z' = 1.33$ and $\mathcal{P}[U \geq U'] \approx 0.03$, that is, about 3 percent.

Thermodynamic fluctuations are often negligibly small, which is why we're only discussing them in this last chapter. Nevertheless, for microscopic systems (e.g., biological cells) fluctuations can drive important molecular mechanisms. Fluctuations are also significant when integrated over large volumes, such as density fluctuations in air causing Rayleigh scattering, which is why the sky is blue.

14.2   Fluctuation Dynamics

Having established a microscopic interpretation for thermodynamic state variables we turn our attention to the phenomenological laws from Chapter 11. Again, consider the heterogeneous, isolated system discussed in the previous section. As before we take $M_{\text{L}} \gg M_{\text{R}}$ so the Left subsystem plays the role of a reservoir with fixed temperature T_{eq}. Macroscopically we have a phenomenological law that relates the heat flux to the temperature difference. We'll write this as Newton's law of cooling

$$\frac{d}{dt} U_{\text{R}} = \mathfrak{C}\, (T_{\text{L}} - T_{\text{R}}) = \mathfrak{C}\, (T_{\text{eq}} - T_{\text{R}})$$

$$= \Lambda\, (U_{\text{eq}} - U_{\text{R}})$$

where $\mathfrak{C}$ is the coefficient for heat conduction and $\Lambda = \mathfrak{C}/c_V M_{\text{R}}$. Since $\delta U_{\text{R}} = U_{\text{R}} - U^{\text{eq}}$ then

$$\frac{d}{dt} \delta U_{\text{R}} = -\Lambda\, \delta U_{\text{R}},$$

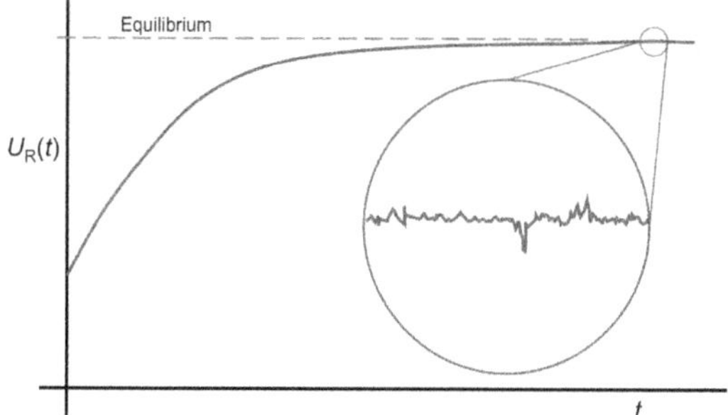

Figure 14.2: Relaxation of $U_R(t)$ towards equilibrium is smooth at macroscopic scales but we see fluctuations when zoomed in to microscopic scales.

which would mean that the deviation from equilibrium decays exponentially in time.

Yet this is only true for perturbations occurring at macroscopic scales (see Figure 14.2). We're missing something because U_R should have some random variation. While the average $\langle \delta U_R \rangle = 0$, we saw in Section 14.1 that $\langle \delta U_R^2 \rangle > 0$ near equilibrium. The question now is how do we add a microscopic correction to the macroscopic phenomenological law?

Brownian Motion

To address this question we consider a seemingly unrelated problem: Brownian motion, which is the random movement of a small particle suspended in a fluid (e.g., pollen grain in water). At macroscopic scales a particle with velocity $\mathbf{v}$ experiences a Stokes drag force characterized by a friction coefficient, λ. The equation of motion in the x-direction is

$$\frac{d}{dt} v_x = -\lambda v_x$$

where the right-hand side is the acceleration due to frictional Stokes drag. The equations for the y and z directions are similar.

At microscopic scales the Brownian particle also experiences significant random forces (e.g., impacts with individual water molecules). A better approximation for the motion is given by the *Langevin equation*,

$$\frac{d}{dt} v_x = -\lambda v_x + B\zeta.$$

The random function $\zeta(t)$ varies rapidly around zero, specifically, $\langle \zeta(t) \rangle = 0$ and

$$\langle \zeta(t)\,\zeta(t+\tau) \rangle = \delta_D(\tau) = \lim_{\sigma \to 0} \mathcal{G}(\tau; 0, \sigma)$$

where δ_{D} is the Dirac delta-function. In the frequency domain the power spectrum of ζ is flat* so it's called a *white noise*. At equilibrium the average velocity of the Brownian particle is zero, that is, $\langle \mathbf{v} \rangle = 0$, but the average speed is *not* zero. From the theory of stochastic processes[†]

$$\langle v_x^2 \rangle = \frac{B^2}{2\lambda},$$

which links the variance of velocity with B, the magnitude of the white noise term.

The equipartition theorem from statistical mechanics tells us that the kinetic energy of a Brownian particle is the same as that of a single, monatomic ideal gas particle

$$\tfrac{1}{2} m \langle |\mathbf{v}|^2 \rangle = c_V m T_{\mathrm{eq}} = \tfrac{3}{2} k_{\mathrm{B}} T_{\mathrm{eq}}.$$

By symmetry $\langle v_x^2 \rangle = \tfrac{1}{3} \langle |\mathbf{v}|^2 \rangle$ so

$$B = \sqrt{2\lambda \langle v_x^2 \rangle} = \sqrt{2\lambda k_{\mathrm{B}} T_{\mathrm{eq}}/m}.$$

Interestingly, the noise term B depends on the friction coefficient λ, which is an example of a *fluctuation-dissipation relation*.

Stochastic Phenomenological Laws

Let's return to our thermodynamic system. Following the same line of thought we now write the phenomenological law as

$$\frac{d}{dt} \delta U_{\mathrm{R}} = -\Lambda \delta U_{\mathrm{R}} + \mathcal{B}\zeta.$$

This Langevin equation is similar to that of Brownian motion. As with Brownian motion the noise coefficient is given by

$$\mathcal{B} = \sqrt{2\Lambda \langle \delta U_{\mathrm{R}}^2 \rangle} = \sqrt{2\Lambda k_{\mathrm{B}} c_V M_{\mathrm{R}} T_{\mathrm{eq}}^2}$$

using the result for $\langle \delta U_{\mathrm{R}}^2 \rangle$ from Section 14.1.

We can now formulate a stochastic version for the dynamics that includes microscopic fluctuations. To do so we write the thermodynamic flux as

$$
\begin{aligned}
\mathcal{J}_U = \frac{dU_{\rightarrow}}{dt} = \frac{d}{dt} \delta U_{\mathrm{R}} &= -\Lambda \delta U_{\mathrm{R}} + \mathcal{B}\zeta \\
&= \mathcal{J}_U^{\mathrm{mac}} + \mathcal{J}_U^{\mathrm{mic}}
\end{aligned}
$$

---

*The Fourier transform of a delta function is a constant.

[†]C. Gardiner, *Stochastic Methods: A Handbook for the Natural and Social Sciences*, Springer (2010).

where $\mathcal{J}_U^{\mathrm{mac}}$ is the macroscopic flux and $\mathcal{J}_U^{\mathrm{mic}}$ is the random flux due to microscopic fluctuations. The thermodynamic force is

$$\mathcal{F}_U = \frac{1}{T_{\mathrm{R}}} - \frac{1}{T_{\mathrm{eq}}} \approx \frac{T_{\mathrm{eq}} - T_{\mathrm{R}}}{T_{\mathrm{eq}}^2}$$

so

$$\mathcal{J}_U^{\mathrm{mac}} = -\Lambda \delta U_{\mathrm{R}} = \mathfrak{C}\,(T_{\mathrm{eq}} - T_{\mathrm{R}}) = L\,\mathcal{F}_U$$

with

$$L = \mathfrak{C}\,T_{\mathrm{eq}}^2 = \Lambda\,c_V M_{\mathrm{R}} T_{\mathrm{eq}}^2$$

being the Onsager coefficient (see Section 11.1).

For the stochastic part of the heat flow

$$\mathcal{J}_U^{\mathrm{mic}} = \mathcal{B}\zeta = \sqrt{2\Lambda \mathrm{k_B} c_V M_{\mathrm{R}} T_{\mathrm{eq}}^2}\,\zeta = \sqrt{2\mathrm{k_B} L}\,\zeta.$$

Collecting the above we have

$$\begin{aligned} \frac{d}{dt}\delta U_{\mathrm{R}} &= L\,\mathcal{F}_U + \sqrt{2\mathrm{k_B} L}\,\zeta \\ &= \mathfrak{C}\,T_{\mathrm{eq}}^2\,\mathcal{F}_U + \sqrt{2\mathrm{k_B}\mathfrak{C}\,T_{\mathrm{eq}}^2}\,\zeta \end{aligned}$$

which is the stochastic version of the phenomenological law for heat conduction. As with Brownian motion we have a fluctuation-dissipation relation since the magnitude of the noise term, which drives the fluctuations, goes as $\sqrt{\mathfrak{C}}$ while the coefficient $\mathfrak{C}$ determines the rate of heat conduction (i.e., dissipation).

This stochastic formulation has practical uses. For example, molecular dynamics computer simulations calculate the intermolecular forces to model physical systems at the microscopic scales. By measuring fluctuations we can determine transport properties, such as thermal conductivity, by applying *Green-Kubo relations* derived from our stochastic phenomenological laws. Surprisingly, the fluctuation data from a simulation that is *in* thermodynamic equilibrium tells us how the system will behave when it's *out* of equilibrium (e.g., what the heat flux will be when there's a temperature gradient).

* * *   * * *   * * *

Every author writes because they have a story they want to tell. This book tells Act I, Scene I of the story of thermodynamics. You've met the protagonist, entropy, and the other main characters. The stage is set and you've seen a glimpse of the future adventures that lie ahead. I encourage you to continue this story and also contribute some lines to it yourself. Have fun!

Acknowledgements Thanks go out to my colleagues at Berkeley Lab and at San Jose State University, both in physics and animation, and a special thanks to all my thermodynamics students over the years.

Image Credits All images are by the author or are in the public domain.

Symbol List

Bold face indicates a vector list, typically over species (e.g., $\boldsymbol{\mu} = \{\mu_1, \mu_2, \ldots\}$). Dot product is used to indicate summation, for example, $\boldsymbol{\mu} \cdot \mathbf{M} = \sum_k \mu_k M_k$. Thermodynamic states are labeled as unitalicised A, B, etc.; for example W_{AB} is the work done on a system going from state A to state B. Heterogeneous systems with Left and Right subsystems have state variables subscripted L and R. Material states (e.g., chemical species) are generically labelled as $\mathbb{A}$, $\mathbb{B}, \ldots, \mathbb{Z}$.

Roman Letters

a Van der Waals coefficient for intermolecular attraction

$\mathfrak{a}$ Generic parameter

A Surface area

$\mathcal{A}$ Chemical affinity

$\mathfrak{A}$ Generic parameter

b Van der Waals coefficient for intermolecular repulsion

$B, \mathcal{B}$ White noise coefficient

c_V, c_p Specific heat capacity at (constant volume), (constant pressure)

$\hat{c}_V, \hat{c}_p$ Molar heat capacity at (constant volume), (constant pressure)

C_V, C_p Heat capacity at (constant volume), (constant pressure)

$\mathfrak{C}$ Heat conduction coefficient in Newton's law of cooling

$\mathcal{C}$ Generic constant or coefficient

$d_\mathrm{i}, d_\mathrm{e}$ Differential change due to (internal),(external) source

f Generic function

f_d Classical degrees of freedom

F Helmholtz free energy

$\mathcal{F}$ Thermodynamic force

$\mathbf{F}_\mathrm{e}$ External mechanical force

$\mathfrak{F}$ Generalized mechanical force

g Acceleration of gravity; As a script, the gas state

G Gibbs free energy

$\mathcal{G}$ Gaussian function

$h; \hbar$ Planck's constant; $h/2\pi$

H Enthalpy

$\mathcal{H}$ Hamiltonian

i Generic index

I Electric current

j Generic index

$\mathcal{J}$ Thermodynamic flow

k, k_0 Generic index, elastic constant

k_B Boltzmann's constant

K Number of material species or phases.

$\mathcal{K}$ Refrigerator coefficient of performance

ℓ System length. As a script, the liquid state

L Onsager phenomenological coefficient (conductivity)

$\mathcal{L}$ Lagrangian

m Mass of a molecule

$M; M_k$ Total system mass; Total mass for species k

n Number density, generic variable

$N; N_k$ Number of moles; Number of moles for species k

N_A Avogadro constant

$\mathcal{N}$ Number of molecules

p Pressure; Hamiltonian momentum

P Onsager phenomenological coefficient (resistance)

q Lagrangian coordinate

$\mathfrak{q}$ Electric charge of a particle

Q Heat added to a system

$\mathcal{Q}$ Electric charge of a system

R Radius; Number of reactions

R Gas constant

$\mathfrak{R}$ Electrical resistance

s Entropy density. As a script, the solid state

S Entropy

t Time or generic parameter

T Temperature

u Internal energy density

U Internal energy

v As a script, the vapor state

V Volume

w Generic variable

W Work done on a system

$\mathcal{W}$ Work done by a system

x Generic variable

$\mathbf{x}$ Spatial location

$\mathcal{X}$ Generalized mechanical displacement

$y; y_k$ Generic variable; Mass fraction for species k

z Generic variable

$\mathfrak{z}$ Fugacity

$\mathcal{Z}$ Partition function for canonical ensemble

$\mathfrak{Z}$ Grand partition function

Greek Letters and Misc. Symbols

α Coefficient of thermal expansion

γ Ratio c_p/c_V

γ_{st} Surface tension

δ Fluctuation from the mean (e.g., $\delta x = x - \langle x \rangle$).

δ_{D} Dirac delta function

Δ Difference (e.g., $\Delta U_{\mathrm{AB}} = U_{\mathrm{B}} - U_{\mathrm{A}}$)

ζ White noise

η Heat engine efficiency

κ_T Isothermal compressibility

λ Drag coefficient in Brownian motion

λ_{T} Thermal wavelength

Λ Thermal conduction dissipation rate

μ, μ_k Specific chemical potential (for species k)

$\hat{\mu}, \hat{\mu}_k$ Molar chemical potential (for species k)

ν Stoichiometric coefficient

ξ Extent of reaction (moles consumed or produced)

π Ratio of circumference to diameter

ρ Mass density

ς Rate of local entropy production

σ Standard deviation

Σ As a script, sum over all subsystems and reservoirs

τ Time increment

ϕ Electric potential

Φ Landau thermodynamic potential

Ω Multiplicity of microscopic states

$[k]$ Molar concentration for species k ($[k] = N_k/V$)

$\langle \cdot \rangle$ Mean (average) value

$\circlearrowright; \circlearrowleft$ Clockwise cycle; Counter-clockwise cycle

CHAPTER 14. THERMODYNAMIC FLUCTUATIONS*

Index